Novel Perspectives of Stem Cell Manufacturing and Therapies

Edited by Diana Kitala and Ana Colette Maurício

Published in London, United Kingdom

IntechOpen

Supporting open minds since 2005

Novel Perspectives of Stem Cell Manufacturing and Therapies
http://dx.doi.org/10.5772/intechopen.82986
Edited by Diana Kitala and Ana Colette Maurício

Contributors
Joana P. Miranda, Ana Serras, Madalena Cipriano, Pedro Silva, Tokiko Nagamura-Inoue, Fumitaka Nagamura, Fabiana Bressan, Lais Vicari De Figueiredo Pessôa, Kaiana Recchia, Naira Caroline Godoy Pieri, TANSU SAYAR KANYIŞ, Ezgi Arslan, Oğuzhan Kanyiş, Dayong Gao, Shen Ren, Shao-Wen Hung, Chia-Chi Chen, Ying-Ching Hung, Chia-Yu Lin, Hsiao-Yun Chen, Ping-Min Huang, Vincent S. Gallicchio, Karley Bates, Diana Kitala, Wojciech Łabuś, Zhiquan Shu, Jiaji Pan, Ji Peng, Junlan Wang, Chunhua Zhao

First published in London, United Kingdom, 2021 by IntechOpen
IntechOpen is the global imprint of INTECHOPEN LIMITED, registered in England and Wales, registration number: 11086078, 5 Princes Gate Court, London, SW7 2QJ, United Kingdom
Printed in Croatia

British Library Cataloguing-in-Publication Data
A catalogue record for this book is available from the British Library

Additional hard and PDF copies can be obtained from orders@intechopen.com

Novel Perspectives of Stem Cell Manufacturing and Therapies
Edited by Diana Kitala and Ana Colette Maurício
p. cm.
Print ISBN 978-1-83962-503-9
Online ISBN 978-1-83962-513-8
eBook (PDF) ISBN 978-1-83962-514-5

For EU product safety concerns:
IN TECH d.o.o., Prolaz Marije Krucifikse Kozulić 3, 51000 Rijeka, Croatia, info@intechopen.com or visit our website at intechopen.com.

Meet the editors

Assistant Professor Diana Paula Kitala graduated with degrees in Biotechnology and Biomedical Engineering, and while writing her doctoral thesis she completed postgraduate studies in the fields of clinical research, biostatistics, laboratory diagnostics, LEAN methodology, and Six Sigma management. She works in a tissue bank as GMP Head of Quality Assurance, in Dr. Stanisław Sakiel Burns Treatment Center, and as a university lecturer. She has taken part in scientific grants, and in 2019 she won the EWMA grant for the project "Theory of constraints (TOC) and LEAN management for wider application of amniotic mesenchymal stem cells in a group of patients with chronic wounds." She was nominated for the golden medal for merits for the Silesian Voivodeship. Personally, she is a mom of two daughters and a volunteer in saving stray animals.

Ana Colette Pereira de Castro Osório Maurício obtained a Veterinary Medicine degree in 1995, a Ph.D. in Veterinary Sciences in 1999 from Faculdade de Medicina Veterinária (FMV) - Universidade de Lisboa (ULisboa), and a Habilitation in Veterinary Sciences from the Institute of Biomedical Sciences Abel Salazar - University of Porto (ICBAS-UP) in 2011. At present, she is an associate professor with Habilitation, from the Veterinary Clinics Department of ICBAS-UP, and scientific coordinator of the Regenerative Medicine and Experimental Surgery sub-unit at Centro de Estudos de Ciência Animal (CECA) from Instituto de Ciências, Tecnologias e Agroambiente da Universidade do Porto (ICETA). For the past twelve years, she has coordinated a multidisciplinary research group of experimental surgery and regenerative medicine, working with several biomaterials and cellular therapies. She started working with embryonic stem cells obtained by somatic nuclear transfer for therapeutic use with Ian Wilmut´s group (Dolly's cloned sheep). More recently, a continued effort to identify and characterize novel stem cell populations appears critical for widespread clinical success. This effort implies in vitro studies, experimental surgery, and in vivo testing, before clinical trials and compassionate treatment in such clinical cases where the traditional and standard treatments have failed. She created a multidisciplinary team, including veterinarians, engineers, and medical doctors that through experimental surgery have a crucial role in the development of biomaterials and cellular therapies. This team allows for sharing of knowledge between biomaterials design, development of cellular systems, and clinical application. She is the supervisor of several Ph.D., post-doctoral, and master's students (sixteen Ph.D. theses already concluded with success and twelve Ph.D. theses ongoing) as well as co-author of many scientific articles and scientific book chapters. She was the principal researcher of several national and international scientific projects and has edited two international scientific books and invented three international patents

Contents

Preface

One of the main goals in medicine is to replace damaged tissues and regain organ functionality, which is hard to obtain by routine therapeutic procedures but might be achieved thanks to stem cells [1]. Lately, stem cell therapy has become a hot topic and as a treatment method has evoked great expectations. There are many books about stem cells, but most of them should be considered out of date when we consider how much new data are presented every year. Other books are extremely expensive, which narrows their readership. Equitable access to information and knowledge about legal aspects, isolation, preparation, and application possibilities may influence the treatment of many diseases in many countries. Equal opportunity for high-quality data about stem cells must include a realistic description of therapies to create a full picture for a critical approach to flooding the amount of material in this topic. Treating autism, rebuilding skin, injecting directly into the brain after a stroke to replace damaged cells, and understanding how diseases occur. Can stem cells really do all these things? What is evidence-based medicine and what is fake therapy and expectations? My entire educational path was devoted to understanding the real potential and worth of stem cell therapy. But firstly – what are stem cells? Stem cells are present both in embryonic and adult cells [2] and are important for newly formed organisms for many reasons. They can develop into many different cell types in the body during growth. Embryonic stem cells can differentiate into every cell type in the body and have unlimited potential for self-renewal, which makes them attractive cell sources for regenerative medicine. However, is the use of these cells ethically acceptable? Are they safe? Are there any other sources of stem cells? Do other stem cell types have similar properties and potentials? How to gather these cells? Are there any protocols? Where are stem cells stored? How can they be used? Are there any legal restrictions? There are many questions to be organized, revised, and summarized. This book is divided into three sections. The first section describes tissue bank institutions and the processing of stem cells. It is crucial to process stem cells in a standardized way because isolating and cultivating stem cells may result in obtaining a heterogeneous population, which does not fit the criteria of the International Society for Cellular Therapy [3]. Although stem cells have great potential in the treatment of many diseases, after their processing, there remain many challenges to overcome before their therapeutic application [4]. The second section addresses the legal and ethical challenges of using stem cells. The third and final section examines clinical use and particular therapies of stem cells. Overall, this special volume presents authoritative views on what we know about stem cells.

Writing this book would not have been possible without Mateo Pulko, Author Service Manager at IntechOpen, who had great patience and provided me with a lot of support. I would like to thank everyone involved in stem cell research, as stem cell therapies have the potential to affect the lives of people all over the world.

Diana, Kitala
Scientific Laboratories,
Tissue Bank,
Dr. Stanisław Sakiel Burn Treatment Center,
Siemianowice Śląskie, Poland

Ana Colette Maurício
University of Porto,
Portugal

References

[1] Zakrzewski, W., Dobrzyński, M., Szymonowicz, M. et al. Stem cells: past, present, and future. Stem Cell Res Ther. 2019, 10(1):68.

[2] Rohban R, Pieber TR. Mesenchymal Stem and Progenitor Cells in Regeneration: Tissue Specificity and Regenerative Potential. Stem Cells Int. 2017:5173732, 2017.

[3] Squillaro T, Peluso G, Galderisi U. Clinical Trials With Mesenchymal Stem Cells: An Update. Cell Transplant. 25(5):829-48, 2016.

[4] Li W, Shi L, Hu B, Hong Y, Zhang H, Li X, Zhang Y. Mesenchymal Stem Cell-Based Therapy for Stroke: Current Understanding and Challenges. Front Cell Neurosci. 15 :628940, 2021.

Section 1

Processing and Preservation Technology

Chapter 1

Introductory Chapter: Stem Cells - Do We Really Know Everything Already?

Diana Kitala and Wojciech Łabuś

1. Introduction

The current trend in medicine divides into two aspects of treatment - preventive medicine and regeneration of damaged tissues (regenerative medicine) [1]. One of goals of regenerative medicine is to improve wound healing without major surgical procedures and donor-site morbidity. This may be obtained by cell-based therapy [2]. The potential to restore tissue to its pre-injured state may be achieved by novel therapeutic modalities, including stem cells. Stem cells have been proposed for the treatment and management of variety of diseases in recent years. In PubMed 366,445 results occur when stem cells phrase is searched, including 5,547 publications describing clinical trials (**Figure 1**).

2. Current knowledge

There is still much to know about stem cells. How they differentiate? Why some therapies may not be effective? How can we use stem cells in other way? Latest studies shown that shape change may be indicator of cell's exit from pluripotency and prevention of release cell membrane from actin cortex may prevent cells from differentiation [3]. It is know that microenvironment impact cells responses which direct tissue growth and physiology [4]. Influencing cell responsiveness may

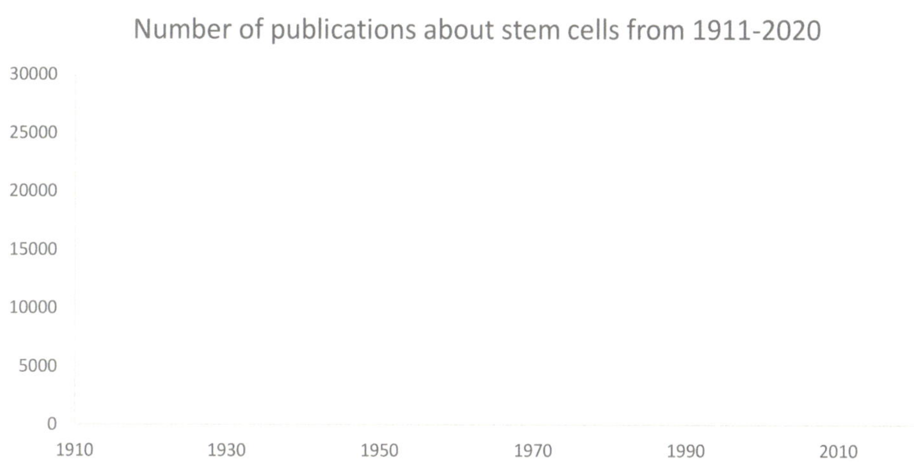

Figure 1.
Stem cell studies and research development during over 100 years.

enhance the regenerative potential of stem cells and therefore improve therapies. Combination of mechanobiological factors can support MSCs to promote vascular regeneration, enhancing endogenous regeneration differentiating them to cell types crutial in neoangiogenesis [5]. But may stem cells be answer in whole organs regeneration or transplantation? Organoids, which are multicellular structures consisting of organ-specific cells are widely created not only to understand organogenesis and disease progression but also in order to potentially treat organ failure [6]. Unfortunately obtaining full organ from stem cells will be impossible for next few decades as organoids made from stem cells are highly simplified [7]. However in other clinical fields stem cells therapies has been widely used for many years.

3. The source

The sources of stem cells for different therapies may vary. Different types of stem and progenitor cells can be isolated from different perinatal tissues, making them particularly interesting candidates for use in cell therapy and regenerative medicine. The main source of perinatal stem cells is cord blood. Cord blood has been a well-known source of hematopoietic stem/progenitor cells since 1974. For over 30 years, umbilical cord blood obtained from biobanks has been used in the treatment of various hematological and immune disorders. Other perinatal tissues that are routinely discarded as medical waste contain non-hematopoietic cells of potential therapeutic value. Indeed, mesenchymal stromal cells are most commonly used in advanced research into perinatal cell therapy [8]. In this perspective, the placenta may deserve special attention [9]. The placenta is a temporary organ that is ejected after birth and is one of the most promising sources of various cells and tissues for use in regenerative medicine and tissue engineering, both experimental and clinical. The placenta has unique intrinsic features because it plays many roles during pregnancy: it is formed by the cells of two people (mother and fetus), contributes to the development and growth of an allogeneic fetus, and has two independent and interacting circulatory systems [8, 9]. The application of MSCs isolated from the amniotic membrane on the acelular dermal matrix ADM allowed for complete healing of extensive wounds of the burned patient without the need to perform an autologous skin transplant [10] which may be a game changer in burn treatment therapy.

4. Clinical applicability

Stem cells are used not only to support wound healing. Stem cells transplantation in patients with multiple sclerosis and amyotrophic lateral sclerosis was a clinically feasible and relatively safe procedure that produces an immediate immunomodulatory effect [11]. Interesting results for the treatment of autism were presented in an open-label Phase I trial. Ongoing inflammation of the nervous system may contribute to the symptoms of autism spectrum disorder (ASD). Mesenchymal stromal cells (MSCs) have been shown to modulate inflammation in the nervous system [12]. Another interesting application of MSCs may be in the future the use of their potential to modulate the immune response during immunosuppressive therapy. Stem cells have been shown to modulate the action of immunosuppressants, and when combined with immunosuppressants, they have a pronounced effect on cell activation and the balance between different T cell subpopulations and exert an inhibitory effect on pro-inflammatory T cell subsets while promoting anti-inflammatory Treg cell function. MSC-based therapy has been shown to be a powerful strategy to mitigate the negative effects of immunosuppressive drugs on the immune system [13]. Those therapies may cause concern about safety but many clinical trials has been studying the safeness and efficacy of mesenchymal stem cells therapies. Most recent describes Phase 1 trial for treatment of COVID-19 patients with pulmonary fibrosis using hESC-IMRCs [14] in which none of the treated patients suffered any adverse events or abnormal responses related to cell therapy. This suggest that stem cell therapies has become more and more safe, popular and researchers discover every year new potential fields of treatment with usage of stem cells.

5. Possibilities

Where else stem cells can be used if not for transplantation? Another option is to study evolutionarily conserved mechanisms with usage of interspecies pluripotent stem cell (PSC) co-culture. This mode enables uncovered a unknown competitive cell interaction during early mammalian development [15].

Stem cells show ourselves many possibilities and when we start think that we know already something a whole new branch of research opens thanks to this unique cells. This book guides us through stem cell types, therapy possibilities, law and ethical aspects. It is complete source of knowledge which is base of further research.

Author details

Diana Kitala* and Wojciech Łabuś
Dr Stanislaw Sakiel Burn Treatment Center, Siemianowice Slaskie, Poland

*Address all correspondence to: dp.kitala@gmail.com

References

[1] Kitala D, Klama-Baryła A, Łabuś W, Misiuga M, Nowak M, Kawecki M. Advanced Technologies in Dermatology. Book: Dermatologic Surgery and Procedures. Edited by Pierre Vereecken. Published: February 28th 2018. ISBN: 978-953-51-3852-5

[2] Domaszewska-Szostek A, Krzyżanowska M, Siemionow M. Cell-Based Therapies for Chronic Wounds Tested in Clinical Studies: Review. Ann Plast Surg. 2019;83(6):e96-e109

[3] Bergert M Lembo S, Sharma S, Russo L, Milovanović D, Gretarsson KH, Börmel M, Neveu PA, Hackett JA, Petsalaki E, Diz-Muñoz A. *Cell Surface Mechanics Gate Embryonic Stem Cell Differentiation.* 2020, Cell Stem Cell, 12:S1934-5909(20)30533-6.

[4] Simmons CA Ireland RG. *Stretch-boosted cell-mediated vascularization.* 2021, Nat Biomed Eng, 5:6-7.

[5] Lee J Henderson, K, Massidda MW et al. *Mechanobiological conditioning of mesenchymal stem cells for enhanced vascular regeneration.* 2021, Nat Biomed Eng, 5:89-102.

[6] Editorial. *The promise of organoids and embryoids.* 2021, Nat Mater, str. 20(2):121.

[7] Rivron NC Frias-Aldeguer J, Vrij EJ, Boisset JC, Korving J, Vivié J, Truckenmüller RK, van Oudenaarden A, van Blitterswijk CA, Geijsen N. *Blastocyst-like structures generated solely from stem cells.* 2018, Nature, 557(7703):106-111.

[8] Torre P Flores AI. *Current Status and Future Prospects of Perinatal Stem Cells.* 2020, Genes (Basel), 12(1):E6

[9] Kitala D Klama-Baryła A, Misiuga M, Łabuś W, Kraut M, Szapski M, Lesiak M, Krakowian D, Sieroń AL, Łos MJ, Kucharzewski M. *Heterogeneous Mixture of Amniotic Cells is Likely a Better Source of Stem Cells than Adipose Tissue.* 2020, Arch Immunol Ther Exp (Warsz).

[10] Kitala D Łabuś W, Klama-Baryła A, Kraut M, Maj M, Szapski M. *Application of Amniotic Stem Cells on an Acellular Dermal Matrix Scaffold in a Burned Patient: A Case Report.* 2020, Transplant Proc, 52(8):2563-2569.

[11] Karussis D Karageorgiou C, Vaknin-DembinskyA, Gowda-KurkalliB, Gomori JM, Kassis I, Bulte JW, Petrou P, Ben-Hur T, Abramsky O, Slavin S. *Safety and immunological effects of mesenchymal stem cell transplantation in patients with multiple sclerosis and amyotrophic lateral sclerosis.* 2010, Arch Neurol, 67(10):1187-1194.

[12] Sun JM Dawson G, Franz L, Howard J, McLaughlin C, Kistler B, Waters-Pick B, Meadows N, Troy J, Kurtzberg J. *Infusion of human umbilical cord tissue mesenchymal stromal cells in children with autism spectrum disorder.* 2020, Stem Cells Transl Med, 9(10):1137-1146.

[13] Hajkova M Hermankova B, Javorkova E, Bohacova P, Zajicova A, Holan V, Krulova M. *Mesenchymal Stem Cells Attenuate the Adverse Effects of Immunosuppressive Drugs on Distinct T Cell Subopulations.* 2017, Stem Cell Rev Rep, 13(1):104-115.

[14] Wu J Zhou X, Tan Y, Wang L, Li T, Li Z, Gao T, Fan J, Guo B, Li W, Hao J, Wang X, Hu B. *Phase 1 trial for treatment of COVID-19 patients with pulmonary fibrosis using hESC-IMRCs.* 2020, Cell Prolif, 53(12):e12944.

[15] Zheng C. Hu Y., Sakurai M. et al. *Cell Competition Constitutes a Barrier for Interspecies Chimerism.* 2021, Nature, strony https://doi.org/10.1038/s41586-021-03273-0.

Chapter 2

Development of a Novel Electromagnetic Rewarming Technology for the Cryopreservation of Stem Cells with Large Volume

Shen Ren, Zhiquan Shu, Jiaji Pan, Ji Peng, Junlan Wang, Chunhua Zhao and Dayong Gao

Abstract

Applications of stem cells have been playing significant roles in scientific and clinical settings in the last few decades. The foundation of these approaches is successful cryopreservation of stem cells for future use. However, so far we can only cryopreserve stem cell suspension of small volumes in the order of 1 mL mostly due to the lack of an effective rewarming technique. Rapid and uniform rewarming has been approved to be beneficial, and sometimes, indispensable for the survival of cryopreserved stem cells, inhibiting ice recrystallization or devitrification. Unfortunately, the conventional water bath thawing method failed in providing the rapid and uniform rewarming. The conversion of electromagnetic (EM) energy into heat provides a possible solution to this problem. This chapter will focus on (1) analysis of the combined EM and heat transfer phenomenon in the rewarming of a biospecimen, (2) numerical investigation of the rewarming system, (3) practical setup of an EM resonance system, and (4) test of heating performance with large volume of cells.

Keywords: stem cell, cryopreservation, large samples, electromagnetic, rapid-uniform rewarming

1. Introduction

Cryopreservation is one the most essential techniques that has been widely used for preserving stem cells in scientific research and cellular therapies [1]. The principle of cryopreservation is to use the super low temperature to reduce the biological and chemical reactions in living stem cells. The expansion in clinical tests for biomedical applications revealed the limitations of the current preservation technologies, i.e., only the small volumes of stem cells can be successfully cryopreserved. In the case of large samples such as bulk volume of cells, tissues, or organs, cryopreservation often fails because of the damage caused by ice crystal growth and thermal stress within the bio-samples [2, 3].

In 1960s, Mazur proposed "two-factor hypothesis" of freezing injury based on the study of Chinese hamster tissue [4]. During the cooling process, two distinct types of cryoinjuries determine the life or death of the cells, which are affected by the cooling rate. Neither too high nor too low the cooling rate is favored for the cryopreservation (**Figure 1**).

When cells are cooled down to subzero temperatures under a normal pressure, ice crystals emerge and grow in the suspensions outside the cell membrane in the beginning [5]. The external ice growth into cells is blocked by the plasma membrane. The cytoplasmic region remains unfrozen and in the supercooled state [6]. However, the increase of osmolality and chemical potential difference across cell membranes due to the external ice formation will serve as a driving force for mass transfer between the intracellular components and extracellular environment, pulling water out of cells. If the cooling is too rapid, there is insufficient time for the water to flow out of the cells. As the temperature goes down rapidly, the unfrozen and supercooled state is disturbed and the intracellular ice formation (IIF) happens [7]. The lethal IIF can rupture the cell membranes and lead to the cell death. On the other hand, if the cooling rate is too slow, intracellular ice may be reduced or avoided. The plenty of time permits water transport out of the membrane under the influence of the osmolality difference. Cells then suffer from high concentration of intracellular solute/electrolytes (so called "solution effects") and severe dehydration. The slower cooling process may expose the cells to "solution effects" for a longer time, which is unacceptable for the cells. Therefore, the cooling rate may not be too high or too low based on the "two-factor hypothesis".

Later in 1984, Mazur reported the rapid rewarming can 'rescue' the rapidly frozen cells [6]. Though the fast cooling process produces intracellular ice, the crystals tend to be small. Due to the unstable thermal properties, during rewarming, small crystals formed at lower temperature aggregate to become larger crystals. The process referred to as recrystallization. It has been proved that the cells cooled at rate far beyond the optimal one will survive if warmed at very rapid rate, but cells do not survive if warmed slowly [8, 9]. Besides the rapid rewarming rate, uniformity of temperature distribution also plays an important role to the survival of the samples [10]. Thermal stress caused by the temperature gradients will lead to the crack of the brittle material, especially the larger systems [11]. Thus, both of fast and homogeneity are essential in the thawing. A rapid-uniform

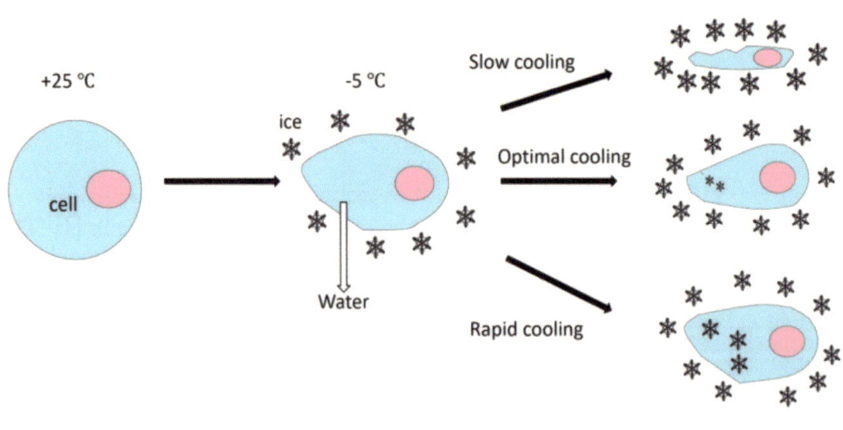

Figure 1.
Schematic drawing of physical events in cells during freezing.

rewarming technology is needed to successful cryopreservation of stem cells with large volume.

A typical cryopreservation procedure is composed of five steps:

1. addition of appropriate cryoprotective agents (CPAs),

2. cooling at an optimal rate,

3. storage at the low temperature (e.g. -80°C freezer or – 196°C liquid nitrogen tank),

4. rearming to physiological temperature, and

5. removal of the CPAs.

Over the past decades, scientists made significant progress in the methods and protocols of addition/removal of CPAs, approaches of cooling to the target temperature, and strategies to preserve biomaterials at sub-zero temperature environment. However, in the scientific and clinical applications, convective rewarming of the cryopreserved samples, typically thawing in a 37°C water bath, remains the gold standard for small samples [12].

2. Rewarming concerns

Warming in the water bath is a convective rewarming approach in which heat is transferred from the outer boundaries to the inner portion of the cryopreserved biological samples. Because of the high specific heat of biological materials, it requires a great amount of heat to rewarm them. Another obstacle for the rewarming process is the low thermal conductivity of biomaterials. Heat cannot be quickly transmitted into the core area of large samples. For a small volume of cell suspensions, the problem caused by the lower thermal conductivity may possibly be solved by the design of sample holder (e.g. maximize the sample holder's ratio of surface area to volume) [13]. However, for larger system with much smaller ratio of the surface area to the volume, they cannot be sliced or pressed into a thin film to increase the heat transfer area for convective warming in water bath. In this case, a large temperature difference will occur and lead to thermal stresses that can result in fracture of the samples during rewarming. Therefore, the traditional method of convective rewarming at the sample surface is not appropriate for the cryopreservation of large volume of biomaterials. A uniform and rapid rewarming method is needed to meet the urgent needs in tissue engineering and cellular therapy, which may be achieved by a volumetric rewarming technique, e.g., electromagnetic heating.

3. Electromagnetic rewarming

3.1 Fundamentals of electromagnetics

The electromagnetic field is a combination of the electric field and the magnetic field generated by electrically charged objects. The electric field and magnetic field are coupled with each other. A time varying electric field (contrary to static field) induces magnetic field changing over time, and vice versa. The distribution and propagation of electromagnetic fields are governed by Maxwell's equations. An electromagnetic wave is the propagation of the electromagnetic field through media such as air, water, etc., or in vacuum. This propagation also refers to radiation transmitting the electromagnetic energy, momentum and angular momentum through space. As one of the basic forces in nature, the propagating electromagnetic fields interact with other materials and generate different effects depending on the frequency and power. Based on the mechanisms behind these interactions between electromagnetic waves and other materials, electromagnetic waves have been applied in numerous aspects including: telecommunications such as mobile phone calling, the short wave broadcast, TV signal transmission and connection between the spaceship and the base on earth, remote sensing for the weather forecasting, land mapping, infrared radiation detecting, X-rays and computer tomography (CT) or the disease diagnoses, and radiotherapy which employs higher energy radiation to kill cancer cells. These various applications are closely related to the spectrum of electromagnetic waves. The most commonly used frequency band is at radio frequency (RF) electromagnetic wave and microwave.

3.2 Electromagnetic rewarming

The thermal effects of electromagnetic waves benefit the organism on earth even long before the prehistory. The thermal energy transmitted by the electromagnetic waves emitted from the sun allows the survival of plants, animals. This electromagnetic wave emitted from the sun, or sunlight, is distributed across almost the entire spectrum but with major intensity on infrared, visible, and ultraviolet frequency ranges. These major components visible light has a frequency range in several hundreds of THz. Higher frequency electromagnetic waves such as X-rays carry much higher energy and are used for radiation therapy killing the tumor cells by ionizing the molecules. In the lower frequency range, electromagnetic waves (radio frequency electromagnetic waves, microwave) with less energy can hardly excite ionization, but can result in thermal effects with materials.

The investigation of the heating phenomenon by microwave occurred in the 1940s. A candy bar under an active microwave generating devices melted and the engineer used this observation to develop equipment for preparing and heating food utilizing the thermal effects of microwave radiation. Since then, not only has the commercialization of the microwave oven become widespread, scientists also began to study the potential benefits for industry and medicine. Most of food is a dielectric material that can interact with the electromagnetic fields in the microwave oven due to a large portion of water in the contents. The water molecules are dipole molecules comprising an electrical positive charged end and an electrical negative charged end. Under the influence of electric fields, these dipole molecules will align themselves with the applied electric fields. When the directions of external electric fields are changing as in the oscillating electromagnetic waves, the dipole molecules will rotate to follow up with the changing fields. The interactions between the rapid changing electromagnetic fields and water dipole molecules will cause mechanical friction forces among the water molecules. The frictions between these micro molecules generate heat volumetrically. Similarly, to the food in the microwave oven for heating, the biological materials in cryopreservation are organisms that are primarily composed of water. Therefore, microwave can also be used for the heating of cryopreserved biological systems.

When using microwave as the rewarming approach in the cryopreservation, the biomaterials interact with the applied electromagnetic field. Since most biomaterials are nonmagnetic, the forced movement of molecules is mainly due to the electric field component of the electromagnetic field. As shown in **Figure 2**, a distinct

DOI: http://dx.doi.org/10.5772/intechopen.94556

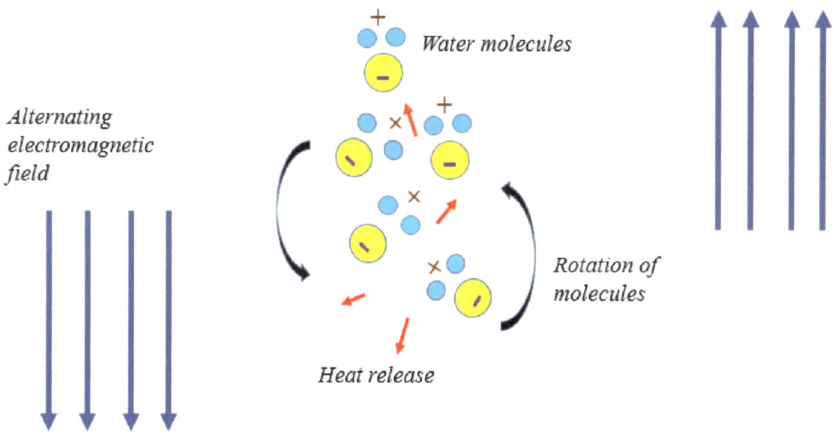

Figure 2.
Illustration of heat generation under the influence of electromagnetic waves.

advantage of microwave rewarming compared with the traditional water bath rewarming is that the heat is generated over the entire region of the material which can lead to volumetric and uniform heating.

The electrical parameter that characterizes the interaction between the material and electric field component is the relative complex permittivity, or dielectric property, expressed in Eq. (1):

$$\varepsilon_r = \varepsilon' - i\varepsilon'' \ . \tag{1}$$

The real part of complex permittivity, ε', also known as dielectric constant, represents the ability of storing electric field energy. The imaginary part of complex permittivity, ε'', known as dielectric loss, represents the ability to absorb electric field energy. The heat generation density q due to the electromagnetic rewarming in the material is given by Eq. (2):

$$q = \pi f \varepsilon_0 \varepsilon'' |E|^2 , \tag{2}$$

where f is the frequency of electromagnetic wave, ε_0 is the electric permittivity of free space, $|E|$ is the root mean square magnitude of the electric field. To deal with the slow warming obstacle due to the high heat capacity, two straightforward solutions can be figured out based on the heat generation principle. One is to select material with higher dielectric loss, and the other approach is to increase the magnitude of the electric field applied to the cryopreserved biomaterial. Provided a strong electric field intensity, the rewarming of cryopreserved biomaterials by electromagnetic wave could be ultrafast which seems to be a promising approach to avoid recrystallization and/or devitrification.

3.3 Previous electromagnetic rewarming attempts by Cryobiologists

The development of electromagnetic rewarming systems is limited by the cost and inadequate theoretical guidance. The establishment of an electromagnetic resonance rewarming system involves the selection of frequency source and power, manufacturing of the resonance chamber, optimization of the electromagnetic energy feeding approach to the cryopreserved materials. Setting up a specific resonance rewarming system can take a few months to years and requires substantial funding support. When the system is going to be scaled up to materials of a larger dimension, various parameters of the system should be replaced which will extend the time required for the optimization. The first experimental investigation of using electromagnetic energy in cryopreservation began in the 1970s. In Kettrer's experiment [14], 20 kidneys rewarmed by a microwave generation device with power control were considered partial success. By using a commercial 1.35 kW Toshiba microwave oven which generated by 2.45 GHz magnetron, Guttman [15] reported the electromagnetic rewarming of 16 cryopreserved canine kidneys. Half of the dogs receiving transplantation of these kidneys survived months.

However, in Pegg's attempt [16] to repeat the rewarming of dog kidneys with commercial microwave oven, none of the post-thawing dog kidneys function properly.

Another device designed by Burdette [17] generated an electromagnetic field with an open electromagnetic illumination system. The frequency can be adjusted to several distinct values. The rewarming results of rabbit and canine kidneys were published without the following viability analysis.

These preceding explorations opened a new avenue for cryobiologists, most specialized in biomechanical and biochemical physics, to implement new technologies from electrical engineering in the application of cryopreservation. A major problem for these early investigations of electromagnetic rewarming is that the frequency of the commercial microwave oven is too high to penetrate into the inner part of the cryopreserved biomaterials. A good uniformity should be achieved with lower frequency electromagnetic waves. In addition, electromagnetic rewarming can result in a 'thermal runaway' problem because the dielectric loss of biomaterials generally increases with temperature during rewarming process, which leads to an increasing temperature difference across the sample volume.

Thermal runaway phenomenon is depicted in **Figure 3**. Due to the complex interactions between the material and the applied electromagnetic wave, and non-uniformity of material properties, the temperature distribution may not be that uniform initially. The slightly warmer area of the biological samples has a higher ability to absorb energy from electromagnetic waves and convert into heat than colder area. The temperature difference between these components will be magnified. When the difference in governing properties is significant and favors the ultrafast thawing of the hot area, the nonuniform temperature gradient is even intensified.

To avoid the localized warming associated with 'thermal runaway' and limited penetration depth for the 2.45 GHz high frequency electromagnetic waves, more delicate controlled electromagnetic rewarming systems operating in lower frequencies are required for cryopreservation. Thereafter, a rewarming system using helical coil to generate tens of MHz electromagnetic waves was used to rewarm CPA solutions as a preliminary trial [18]. But the rewarming rate was moderate achieved by this open system. Later on, a few scientists reconsidered the electromagnetic

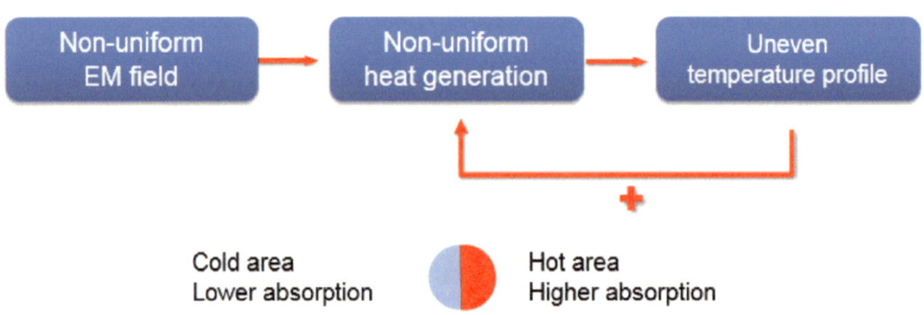

Figure 3.
Introduction of the 'thermal runaway' phenomena: The appearing of large temperature gradient by electromagnetic rewarming.

rewarming in closed systems to confine electromagnetic energies [19, 20]. Unlike the commercial microwave oven in which the generated signal shifted around ±50 MHz, the electromagnetic wave was synthesized by voltage-controlled oscillators. The stability of frequency was improved so that the design of chamber could fit with the electromagnetic source and establish resonant state to concentrate the electromagnetic energy.

A major problem for these resonant systems may come from an intention to reduce the non-uniformity by adopting multimode resonant cavities. In Rachman's electromagnetic resonance cavity rewarming system [19], two resonant states of the cavity were excited (TE111 and TM010). Another multimode cavity rewarming system designed by Robinson [20] excited three different modes. The results of the warming test for CPA solutions were improved while the spatial temperature difference was not greatly reduced, which means thermal runaway could not be eliminated. The reason behind this is due to the interaction between the cryopreserved CPAs and the properties of the resonant cavity. Multimode cavities resonating at different frequencies bring more difficulties in the control of signal frequency at each port to feed electromagnetic energy into the cavity. Hence, a single mode resonant cavity which excites only at a specific frequency may be superior in the control of field distribution in the rewarming process.

Our goal is to achieve effective cryopreservation protocol for bulk volume of stem cells. In this chapter, the target is to improve electromagnetic rewarming systems for the rapid-uniform rewarming, laying down the foundation for the cryopreservation of biomaterials with large volume.

4. Characterization of essential physical properties of CPA solutions

4.1 Determination of the electrical properties

The dielectric properties of the biomaterials characterize the interaction between applied electromagnetic field and the biomaterials, and thereby determine the absorption of electromagnetic energy by the biomaterials [21, 22]. The dielectric properties are temperature dependent. If the warmer part of the biomaterial absorbs more heat, the temperature at that warmer part would be further increased, increasing temperature gradients, and therefore inducing thermal stresses. A large thermal stress can destroy the viability of cryopreserved materials and can be even more threatening to bulky systems [23, 24]. Since the dielectric properties play a key role in the absorption of electromagnetic energy, it is a priority to discover the dielectric properties of the biomaterials. In cryopreservation, particularly in vitrification using a high concentration cryoprotective agent, the CPA/vitrification solutions dominate the properties of the cell suspensions or tissues. Hence, the dielectric properties of the CPA/vitrification solutions should be determined so that electromagnetic rewarming can be optimized.

The measurement of the dielectric properties of biomaterials requires sensing and monitoring tools. In many biomedical applications, various measurement methods including transmission and reflection techniques have been used to determine dielectric properties [25–29].

The samples measured by transmission and reflection methods usually are fixed without morphologically change. But in the application of electromagnetic rewarming, the measurement of dielectric properties must be carried out in the subzero temperature range which may involves phase change and rules out the possibility using transmission and reflection techniques. The cavity perturbation method has been used for measuring the electric properties of different kinds of materials [30–33] due to its ability to measure the dielectric properties of low loss dielectric materials [34]. In the subzero temperature range, the dielectric properties of biomaterials and CPA/vitrification solutions can be very small. Therefore, in this work, we adopted a cavity perturbation method to determine the dielectric properties of three different vitrification solutions at low temperatures. Briefly, a resonant cavity was designed and manufactured to measure the dielectric properties of cryopreserved biomaterials at 434 MHz. By inserting samples with different permittivity into the resonant cavity, the resonant frequency and quality factor could be changed. From the variation of the resonant frequency and the quality factor, the dielectric properties can be derived.

4.1.1 Perturbation theory

Through the change of resonant frequency Δf and inverse of quality factor ΔQ, the dielectric property (or complex permittivity) of CPA/vitrification solutions was obtained. The mathematical derivation is shown by following Equations [35, 36]:

$$\frac{\Delta\tilde{\omega}}{\tilde{\omega}} = \frac{\Delta f}{f_0} + \frac{i}{2}\left(\frac{1}{Q} - \frac{1}{Q_0}\right) \tag{3}$$

$$\frac{\Delta\tilde{\omega}}{\tilde{\omega}} = \frac{\int_{V_s}\left(\Delta\varepsilon E E_0^* + \Delta\mu H H_0^*\right)dv}{\int_{V_c}\left(\varepsilon E E_0^* + \mu H H_0^*\right)dv} \tag{4}$$

where Δf and ΔQ are resonant frequency and quality factor of unperturbed cavity, $\boldsymbol{E}_0$ and $\boldsymbol{H}_0$ represent electric field and magnetic field inside the unperturbed cavity, $\boldsymbol{E}$ and $\boldsymbol{H}$ represent those fields for the cavity with sample to be measured inside, ε and μ are complex permittivity and permeability of the sample, i is an imaginary number of which the square equals -1. More simplification can be applied to Eq. (4), equating the real and imaginary parts and the following relationship will be obtained:

$$\Delta f = k_1 \frac{\varepsilon' - 1}{\varepsilon' + 2} \tag{5}$$

$$\Delta\left(\frac{1}{Q}\right) = k_2 \frac{\varepsilon''}{\left(\varepsilon' + 2\right)^2} \tag{6}$$

k_1 and k_2 are constants to be determined.

4.1.2 Dielectric property measurement system

The experimental system is shown in **Figure 4**. A rectangular single-mode resonant cavity resonating at around 434 MHz was manufactured. The dimension of the cavity was designed to 680 × 400 × 350 mm. Copper plates were used to manufacture the cavity due to its high conductivity (to prevent electromagnetic leakage).

4.1.3 Calibration of the measurement system

Water, methanol, ethanol, 1-propanol, 2-propanol, ethylene glycol and cyclohexane samples with known properties were used for calibration of the cavity to determine k_1 and k_2. The properties of these calibration solutions were reported by Robinson [37].

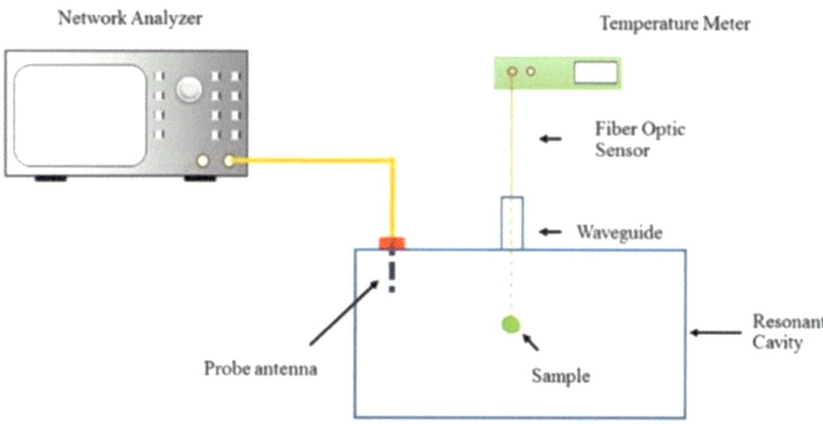

Figure 4.
Schematic of the system for measurement of dielectric properties.

Once k_1 and k_2 had been obtained, unknown cryopreserved biomaterials can be measured by the system. Similarly, the resonant frequency and quality factor were changed. According to the shift frequency and quality factor of the cavity, complex permittivity of samples can be derived.

4.2 Determination of the thermal properties

4.2.1 Measurement of specific heat

To determine the temperature-dependent specific heat capacity, Differential Scanning Calorimetry (DSC) was used to take accurate measurement of various sample solutions. Isothermal step scan method is adopted to minimize the experimental error. The latent heat was incorporated into the effective specific heat capacity when the phase transition occurs.

4.2.2 Measurement of thermal conductivity

Thermal conductivity of sample solutions was measured using a micro thermal sensor developed by Liang, et al. [38] and manufactured in the lab. The sensor works on the principle of Transient Hot Wire (THW). This miniaturized device utilizes a $SiO2/Au/SiO2$ sandwiched structure to protect the microfabricated serpentine gold coil, which functions as both the heater and a passive thermometer. The sensor has already been tested and shown to measure thermal conductivity of biomaterials and solutions with high accuracy, repeatability and system reliability.

5. Theoretic analysis and numerical simulation to improve the electromagnetic resonance system

5.1 Introduction

Previous cryobiologists have developed electromagnetic cavity rewarming for large biomaterials. Evans [39], Robinson [37] and Luo [40] built electromagnetic heating systems that could resonate at around 434 MHz but working at different modes. The sample was placed at the center with a large magnitude of the electric field, which is critical to rapid rewarming. The size of the sample was controlled to reduce temperature gradients resulted from the electromagnetic field attenuated away from the center. For multimode resonant cavity rewarming system, it could be difficult to control the resonant state or field distribution with several EM power inputs. While single mode cavity system is easier to concentrate a strong EM field and control in the rewarming process. Therefore, multimode resonant rewarming systems are excluded for optimization here. According to the warming results, electromagnetic rewarming has already demonstrated more effective than traditional water bath method. However, further improvement of the electromagnetic rewarming system is still needed to optimize the rewarming protocol of bulk volume of cells.

Multiple theoretical analysis and numerical simulations had been accomplished [41–43] on different design possibilities of electromagnetic rewarming systems. Method of moments (MoM) was implemented to calculate and analyze the electric filed intensity and profile [42] and the method of finite-difference time-domain (FDTD) was applied to investigate the post-thawed temperature profile [44]. Another attempt involving the resonance rewarming in the numerical model combined with nanoparticles to improve the warming rate. However, in this study, the model was based on the power consumption at 8000 W [45] that is hard to achieve in the real-world.

In this section, a more efficient and effective model based on finite element method combining electromagnetic wave propagation and heat transfer process was presented. The optimization of the shape of cryopreserved sample was performed numerically. The essential physical properties of several sample solutions including complex permittivity, specific heat, and thermal conductivity characterized experimentally in the previous studies were used in this simulation test.

5.2 Theoretical formulations

The electromagnetic resonant system setup showed in **Figure 5** consists of a signal generator with voltage control at lower scale, a power oscillator, a resonant cavity, a coaxial transmission wire, and a temperature sensor. An antenna made of high electrical conductivity material was used to excite the electric field and create a standing wave pattern of electromagnetic field inside the rewarming cavity. A sample holder contained cryopreserved stem cells or other biomaterials was place in the center of the cavity, where the strongest electric field was established to achieve the fast rewarming.

Most of the cryopreserved biomaterials are insensitive to the magnetic field, e.g. the stem cells, with electrical parameters including, ε (absolute permittivity, $F \cdot m^{-1}$), σ (conductivity, $S \cdot m^{-1}$), and $\mu = \mu_0$ (magnetic permeability of the free space, $H \cdot m^{-1}$,), the frequency domain Maxwell equations were used to calculate the electric field profile in the cryopreserved material:

$$\nabla \times E = -j\omega\mu H - M \tag{7}$$

$$\nabla \times \nabla \times H = -j\omega\varepsilon E + \sigma E + J \tag{8}$$

where E is the electric field intensity ($V \cdot m^{-1}$), and H is the magnetic field intensity ($A \cdot m^{-1}$). Here, M is the magnetization filed intensity ($A \cdot m^{-1}$) that equivalent to the H term, and J is the flowing current density ($A \cdot m^{-2}$) that equivalent to the E term. Both M and J can be ignored since none of magnetic and

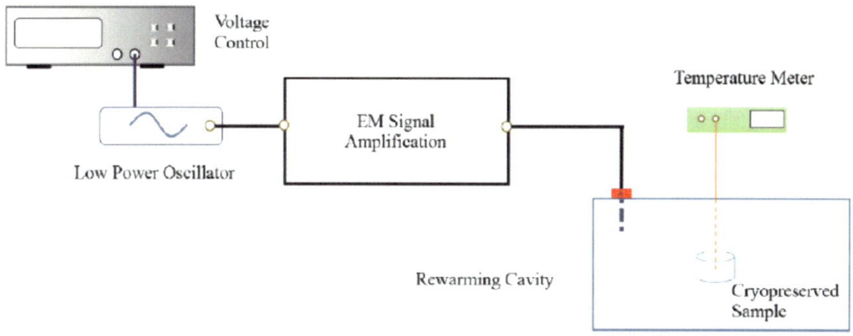

Figure 5.
Schematic description of the experimental setup of the EM rewarming system.

electric source was established inside of the given geometry. ω is the angular frequency of the EM field ($rad \cdot s^{-1}$). The absolute permittivity $\varepsilon = \varepsilon' - i\varepsilon''$, where ε' is the real part, stands for the dielectric constant that describe the space's ability to hold the electric power within the control volume. ε'' is the imaginary part, stands for the dielectric loss that describe the space's ability to harness electric power. Heating is generated through the vibration of dipolar molecules affected by the electric field. The term q represents the absorbed electric power in the material is described as:

$$\nabla \times E = -j\omega\mu H - M \tag{9}$$

$$q = \pi f \varepsilon_0 \varepsilon'' |E|^2 \tag{10}$$

where f (Hz) denotes the frequency of the electromagnetic field. q ($W \cdot m^{-3}$), the total absorbed power from the electromagnetic field, can be combined into the heat transfer equation as the heat source term [46]:

$$\rho C \frac{\partial T}{\partial t} = \nabla \cdot (k \nabla T) + \pi f \varepsilon_0 \varepsilon'' |E|^2 \tag{11}$$

where $\rho\left(kg \cdot m^{-3}\right)$ represents the density of the cryopreserved material, C $\left(J \cdot kg^{-1} \cdot K^{-1}\right)$ denotes the specific heat, k ($W / m \cdot K$) is the thermal conductivity, T (K) is the temperature of the cryopreserved material, and t (s) is the time in the heating process.

In this numerical modeling analysis, the thermal properties of the cryopreserved material were preset to be temperature-dependent. The specific heat and thermal conductivity at different temperatures were captured once the temperature profile in the Eq. (11) was determined. With the updated thermal and electrical properties over the interested temperature range, the electromagnetic field profile was calculated. Cryopreserved material's dielectric loss ε'' and applied electric filed intensity changed the heating source. Thus, the temperature and electric fields are coupled. Adiabatic ($q_{conv} = 0$) assumptions were applied to the boundary conditions to eliminate the natural convection within the cavity. The perfect conductor assumption was also applied to the inner surface of the resonant cavity ($\vec{n} \times E = 0, \vec{n}$ denotes the inner wall's unit vector of the cavity).

5.3 Numerical simulation

The numerical modeling was setup with COMSOL Multiphysics (COMSOL, Burlington, MA, USA), applying the finite element method. Adopting the analytical process discussed in the theoretical formulation section, the thermal science and combined electromagnetic rewarming system in the resonant chamber as shown in **Figure 6**. In this numerical investigation, the thermal and electrical parameters of the model were determined follow the methods discussed in the previous section.

Nyquist criterion was applied to the meshing grid of the simulation. The maximum size of the element grid was contained smaller than half of the wavelength of the electromagnetic wave. Six grids per wavelength in the finite element analysis was also employed to solve the Maxwell's equation [47].

Tetrahedral grids were selected in this study due to smaller grid size comparing to the discretization of the resonant rewarming system. Once the tetrahedron grid applied, the grid size of the resonant chamber was calculated to be less than 10% of the wavelength of electromagnetic wave. In the mesh preparation (shown in **Figure 7**), different meshing approaches were adopted for different components of the system. Refined meshes were created near the boundaries of surfaces (in total, about 988,000 elements were generated). The area around probe antenna

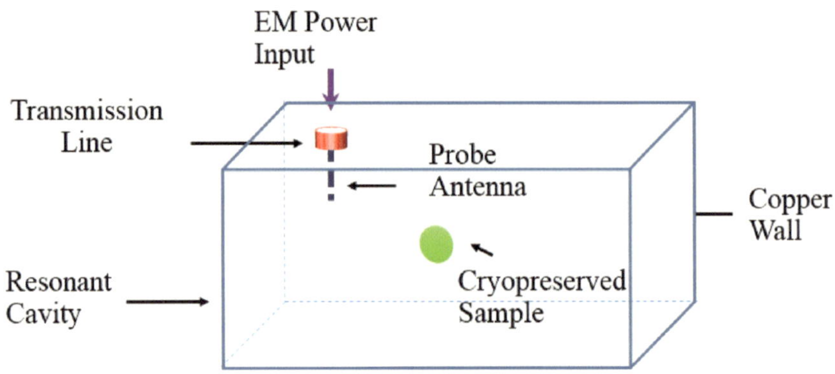

Figure 6.
Schematic description of the simulated resonant EM rewarming cavity.

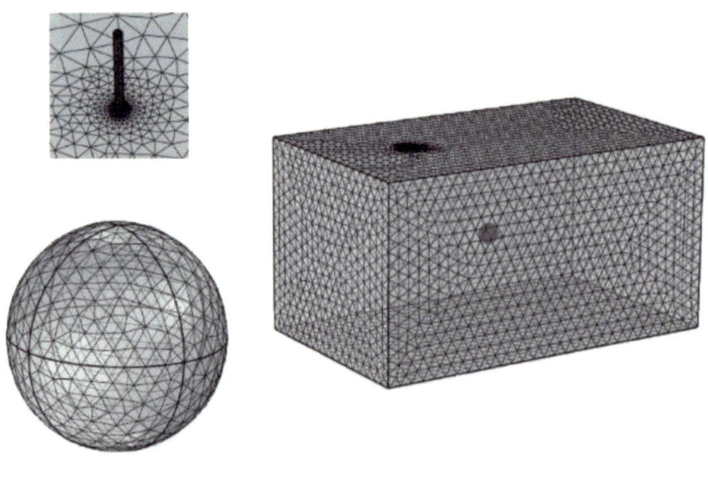

Figure 7.
Geometric gridded model of the resonant cavity for simulation.

and central part of the sample holder were treated with the enhanced meshing. A smaller grid size (about 163,000 elements) will diminish the difference of the temperature profile less than 0.5% over the entire field.

5.4 Results

5.4.1 Field distributions in the resonant cavity

The prepared simulation model was validated with the analytical results. In the model, the resonant frequency of the chamber was 434.767 MHz, consistent with the frequency from the analytical result. When the chamber was at the resonant state, the normalized electric field intensity profile was plotted for both of the simulated result and analytic result in the resonant cavity was plot in comparison with the analytic solution (**Figure 8**), which proofed a good alignment between two results. As the position advancing to the central part of the cavity, the intensity of the electric field increased significantly. Hence, the cryopreserved bio-sample was designed to be placed in the central to achieve a fast rewarming process by absorb more electromagnetic power.

5.4.2 Various sample holder shape's effect on the rewarming process

The effect of sample holder's shape to the rewarming was studied to minimize the injury to the biomaterials caused by non-uniform thawing. On the other side, different sample holders might be selected based on the size and structure of the biomaterials. Considering the hardness to accurate calculate the energy conversion between the cavity and the sample through analytical solution, and precise temperature profile monitoring over the entire sample space through experiment, in this simulation, a total of four different sample holder shapes (geometry details shown in **Table 1**) with the same sample volume of 25 mL were calculated to study the difference over the rewarming process. The heating started at the initial temperature set to −80° C and ended at 0°C. The average rewarming rates of cylindrical, ellipsoidal, spherical, and cubic holders were 72.1, 63.5, 46.1, and 22.8° C/min, respectively. From the fast rewarming perspective, cylindrical and ellipsoidal shapes were the top choices.

Later, the temperature profile at the end of the rewarming was investigated. As shown in **Figure 9**, the temperature gradients that defined as the temperature

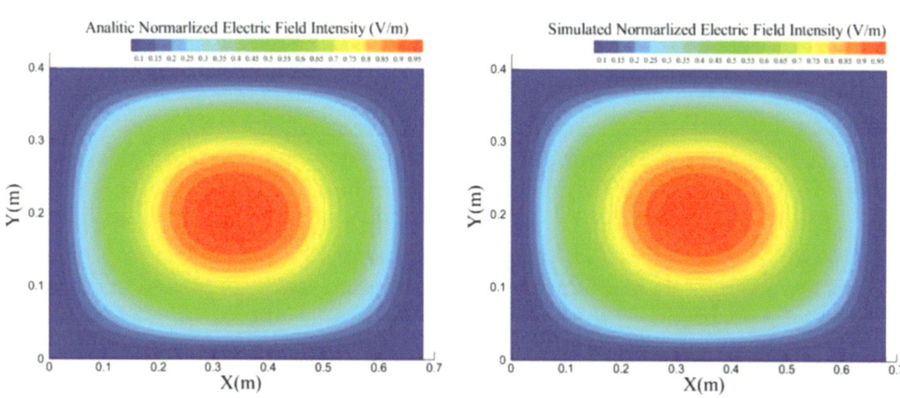

Figure 8.
Distribution of the electrical field magnitude at the central cross-sectional plane in the electromagnetic rewarming chamber. Analytical and simulated results show that electric field energy is focused in the center of the cavity.

Parameter	Value (mm)
Radius of sphere	18
Radius of cylinder	18
Height of cylinder	20
a-Semi axis of ellipsoid	23
b-Semi axis of ellipsoid	14
c-Semi axis of ellipsoid	18
Side length of cube	29

Table 1.
Dimensions of sample holder shapes.

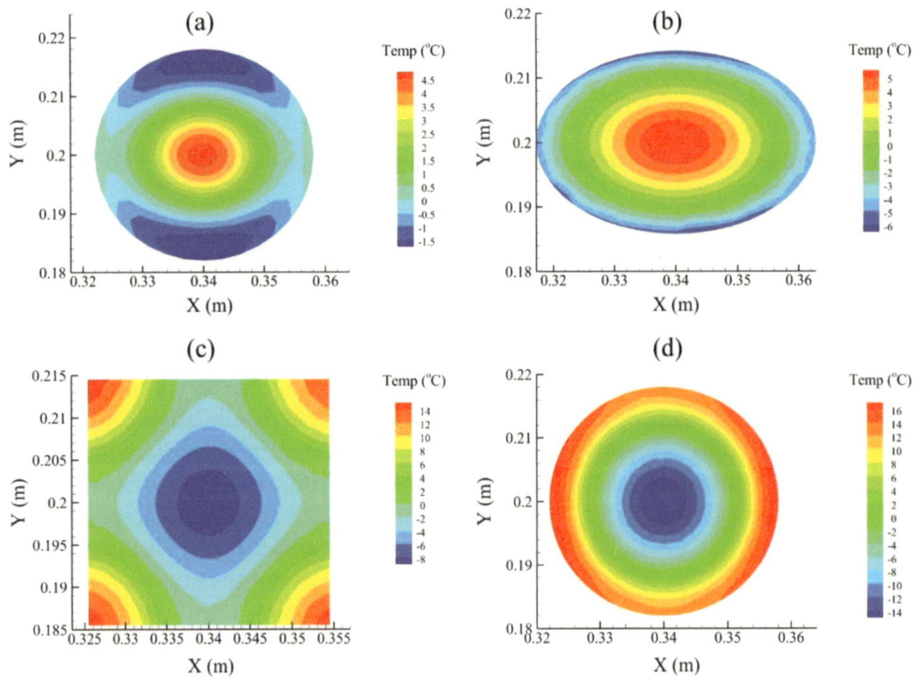

Figure 9.
Simulated post thawing temperature distribution of cryopreserved sample solution in different sample holder shapes. (a) Spherical shape; (b) ellipsoidal shape; (c) cubic shape; (d) cylindrical shape. The spherical and ellipsoidal samples manifested more uniform temperature distribution than cubic and cylindrical samples.

difference between the maximum and minimum temperatures in the holder divided by the ferret diameter, for spherical, ellipsoidal, cylindrical, and cubic samples were 0.27, 0.62, 0.95 and 1.24° C/mm, respectively. From the temperature gradient perspective, the spherical and ellipsoidal shapes were the top choices.

Both rewarming rate and temperature distribution results indicated the cubic shape will not perform well the resonant chamber, which aligned with the prediction that samples with sharp edges or surfaces were inappropriate to be heated by the electromagnetic rewarming method [42].

Robinson et al. presumed that ellipsoidal sample shape works well for electromagnetic rewarming technology and performed the experiment with a cone-shape sample to approximate a ellipsoidal holder [21]. The currents from numerical modeling confirmed the fast and uniform rewarming of the ellipsoidal sample.

However, it is extremally hard to manufacture a precise ellipsoidal shape holder in the real-world. The same hardness also applied to the spherical holder. Moreover, a cryopreserved material should be stored well in an optimized holder to achieve a better rewarming performance. Thus, the material of the holder was ideally a thin-layer to diminish the absorption of the electromagnetic power. Overall, it is an engineering challenge to manufacture a qualified sample holder with desired material and dimensions in spherical or ellipsoidal geometry. Additionally, an extra supporting structure is required to hold the position of the spherical and ellipsoidal sample in the center of the cavity, which may result as the effect of the electromagnetic profile. Therefore, cylindrical holder was the best option with the fair manufacturing and faster rewarming rate. Though, more improvements should be considered to enhance the uniformity of the temperature profile.

6. Real-time resonant frequency monitoring and controlling system

Convective warming methods are hindered by the poor abilities to conduct heat into the core part of the materials. While volumetric heating method is needed for bulky material, the previous multimode or commercial microwave systems could not be adopted since they lack a precise control system to maintain the resonant state, leading to either recrystallization or devitrification. Due to the slow heating, most of electromagnetic energy was reflected back from the cavity or causing thermal runaway problems and creating undesired hot spots.

There are two major limitations for electromagnetic resonance system to achieve rapid and uniform heating. Firstly, the system itself should provide sufficient electromagnetic energy to warm up biomaterials. Secondly, the temperature dependent dielectric properties of the biomaterials progressively shift the resonant frequency of the resonant chamber during the rewarming process. Therefore, when using resonant electromagnetic field as the heating source, if the electromagnetic signal parameters remain stagnant according to the frequency change resulted from the temperature change of biomaterials, the electromagnetic energy generated may not be converted into the strong electromagnetic field in the resonant chamber to excite resonance.

Moreover, if the electromagnetic system source remains static during the rewarming procedure, severe problems regarding the system safety and efficiency may emerge. A higher portion of reflected electromagnetic power can lead to the damage to the system components as well as potential electromagnetic radiation hazards to the surrounding operators. On the other hand, with smaller electromagnetic energy remaining inside the rewarming chamber, sufficiently strong electromagnetic field inside the resonant cavity could hardly be excited resulting a slow warming. Therefore, it requires delicate control on the set up of the electromagnetic resonant system.

6.1 Electromagnetic source

The electromagnetic signal is synthesized by a signal generator (Agilent, Santa Clara, CA, USA). The signal generator can generate continuous electromagnetic waves between 250 kHz and 3 GHz which covers lower frequency band of radiofrequency and microwave. And the power output range is +7 to −120 dBm (i.e. 0.005 to 10–15 W). At this low power output, the electromagnetic field established in the resonant chamber is too weak to rewarm the cryopreserved biomaterials rapid enough avoiding devitrification. In order to intensify the electromagnetic field to achieve higher rewarming rates, a power amplifier (OPHIR RF, Los Angeles, CA, USA) was adopted to increase the electromagnetic power to over 57 dBm (501 W).

The power amplifier has a frequency range between 300 and 500 MHz, which fully covers the working frequency range for this experimental investigation. The reflected power received by this power amplifier would also be detected by the control circuits. The power amplifier will automatically cut off excessive output generation to protect itself. Due to the relatively high electromagnetic power used in the system, two side panel cooling fans were incorporated to avoid internal circuits overheating and system shutdown. The connections between the signal generator and power amplifier, as well as other microwave components are through 50 Ω coaxial cables. These cables would have some attenuation effects. Thus, in order to maintain the high power signal from the amplifier to the rest part of the system, the length should be as short as possible. The measurement was done by the signal generator and a power meter. According to the measurements of six coaxial cables of different lengths, the attenuation for coaxial cables is around 0.1 dB/m.

The most significant difference between the current resonance system and the previous assembled circuits is attributed to the frequency tracking component. During the rewarming process, the resonant frequency of the resonant cavity with biomaterials would change on account of the temperature dependent dielectric properties of the inside biological samples. To prevent the mismatch between the synthesized electromagnetic source frequency and the resonant frequency of the rewarming chamber, the generated frequency source should be dynamically adjusted during the rewarming process.

6.2 Feedback control

In order to prevent the frequency mismatch between the signal generation and the resonant frequency which is swiftly altered by the massive cryopreserved materials inside, a dynamic feedback control component was added between the electromagnetic source and the resonant cavity. A directional coupler was introduced to sample the transmitted and reflected power. A spectrum analyzer was connected to the port corresponding to the reduced reflected power. The entire spectrum of the reflected power was evaluated and by looking for the highest power peak, the frequency corresponding to the most significant reflected power was determined. During the rewarming process, the frequency generated from the electromagnetic signal source is dynamically changed corresponding to this spectrum and minimize the reflected power. Otherwise, the large portion of reflected power could lead to a slow warming rate with less power into the cavity. In addition, the reflected power can cause a serious damage to the rest part of the electromagnetic resonance system itself, such as the amplifier, signal synthesizer.

6.3 Enhancement of electric field magnitude of the resonant cavity

In the numerical simulation model, the probe length was adjusted to be the original probe length and the extended probe length. The electric field intensity excited in the sample inside the cavity was calculated. As shown in **Figure 10**, the electric field intensity in the cryopreserved material increases almost ten times larger than that using the original probe antenna, which suggests that the impedance matching between the loaded cavity and the electromagnetic source is greatly improved by adopting an extended probe antenna.

Since electric field power is proportional to the square of the electric field intensity, we could have much more power to heat the material using the optimized extended probe antenna. This numerical estimation of the electric field gives guidance to the experiment, the reflected power was measured by a network analyzer, and the quality factor of the loaded cavity was determined based on the reflection coefficient. It is

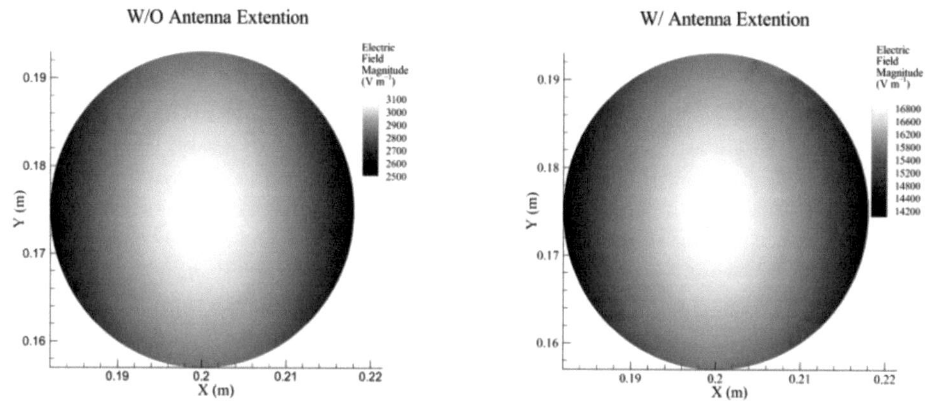

Figure 10.
Electric field intensity comparison.

found the quality factor of the loaded cavity was improved from 1681 to 5577 after adding the probe extension, which can establish a much stronger electromagnetic field inside the cavity for the rapid rewarming of the cryopreserved biomaterials.

7. Test of the electromagnetic resonance system with Jurkat cells

To experimentally evaluate the heating performance of the electromagnetic resonance system, a rewarming test of the cryopreserved Jurkat cells with large volume (25 mL) was performed. Jurkat cell is an easily accessible cell line and shares similar cryopreservation protocol to the stem cells. The testing results will guide us the future trails of cryopreservation of large system of stem cells.

7.1 Materials and methods

7.1.1 CPA solution

The CPA cocktail contains 10% (w/v) dimethyl sulfoxide (Me_2SO ; Sigma-Aldrich, St. Louis, MI, USA) and 0.25 M trehalose (Sigma-Aldrich) in culture medium (Sigma-Aldrich) solution. This CPA combination is widely used for the cryopreservation of stem cells and immune cells. The thermal and electrical properties of CPA solution were determined using methods introduced in Section 4.

7.1.2 Samples for cryopreservation

Human T lymphocyte leukemia cells (Jurkat cells) were used in the rewarming test. The Jurkat cell lines were purchased from American Type Culture Collection (Manassas, VA). Cells were cultured in the incubator that setting at 37°C, 5% carbon dioxide, and proper humidity. T25 flasks were used during the cell culture. A unit of the growth medium was prepared of 450 mL RPMI medium (life technologies), 50 mL fetal bovine serum (FBS), 5 mL Penicillin–streptomycin, and 5 mL L-glutamine.

7.1.3 Cooling process

Cylindrical sample holder contains Jurkat cell suspension was placed in a Styrofoam box, then the box was transferred to the −80°C freezer and stored there overnight. The average cooling rate was 2–3°C/min.

7.1.4 Temperature profile measurement

The temperature measurements were conducted by a fiber optic temperature meter (Neoptix Inc., Ville de Quebec, QC, Canada) during the rewarming process. The major challenge for the temperature measurement lies in the penetration through cavity wall. The cryopreserved sample remains in the center of the resonant chamber where the highest electromagnetic field was formed. However, the penetration through the cavity wall would undermine the quality factor of the resonant cavity, which means lower portion of electromagnetic energy remained inside the chamber for rewarming. Additional waveguide was designed to allow for the fiber optic temperature sensor to get through the chamber wall and maintain the quality factor at the same time. This waveguide was designed to have the cutoff frequency higher than the operating frequency during the rewarming process. Although the side effects associated with the EM waves are still in debate, it is nevertheless safer to keep away from the possible side effects caused by electromagnetic energy leakage.

At the end of the rewarming process, the surface temperature profile was recorded by an infrared temperature sensor (FLIR systems, Wilsonville, Oregon, USA). The temperature data in the central part of the cryopreserved sample recorded by the fiber optic meter and the surface temperature profile are combined to analyze the temperature gradient.

7.2 Results and discussion

7.2.1 Rewarming rate

Figure 11 shows the comparison of rewarming process of Jurkat cells between conventional water bath and electromagnetic resonance system. The average rewarming rate of water bath was 40°C/min, while increased to 90°C/min for the EM system.

7.2.2 Recovery rate of Jurkat cell

Membrane integrity was obtained by Trypan Blue (Sigma-Aldrich) staining to determine the recovery rate of Jurkat cells. **Figure 12** shows the comparison of recovery rate of Jurkat cells between conventional water bath and our

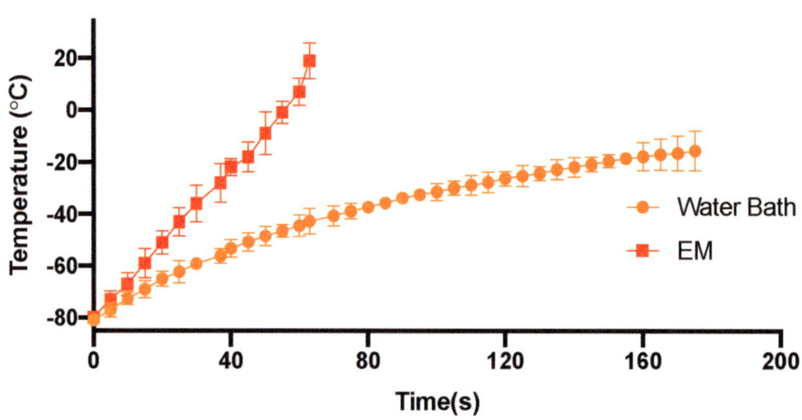

Figure 11.
Rewarming process of Jurkat cell.

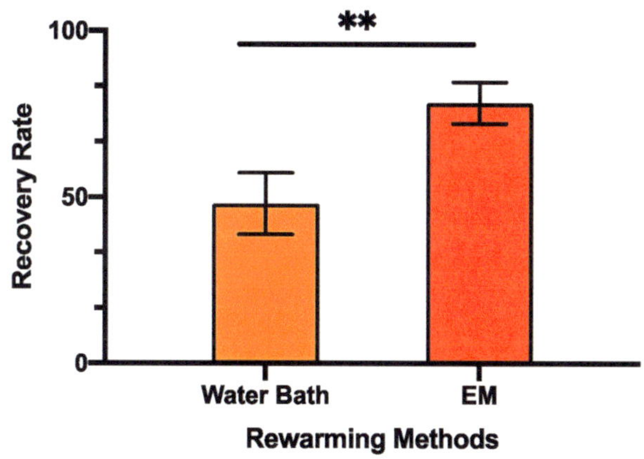

Figure 12.
Recovery rate of the Jurkat cell.

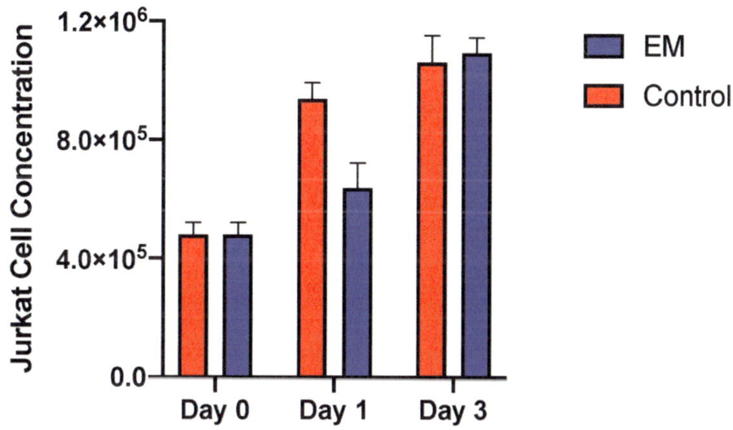

Figure 13.
Proliferation rate of the post-thawed Jurkat cell.

electromagnetic resonance system. Water bath achieved the Jurkat cell's recovery rate at $48.0 \pm 9.3\%$, while EM system significantly improved to $78.3 \pm 6.2\%$. The recovery rate obtained by EM method surpassed water bath by 63.1% ($p < 0.01$).

7.2.3 Post-thawed assessment of Jurkat cells

The post-thawed cell suspensions were cultured in a 37°C incubator with 5% carbon dioxide and proper humidity. As shown in **Figure 13**, after three days incubation, no significant change was noted to the normalized proliferation rate for the electromagnetic rewarming method. This indicates EM system does not affect Jurkat cell's cellular functionality.

8. Conclusion

In this chapter, we provide detailed information about using electromagnetic resonance system to achieve rapid-uniform rewarming in cryopreservation of stem cells. The importance of rapidly and uniformly rewarming process to the bulky system of stem cells was explained, principles of electromagnetic warming were described, essential physical properties of CPA solution and resonance cavity were covered. Theoretical analysis and numerical simulation were introduced to improve the heating performance. A dynamic resonance frequency monitoring and control system was developed. Apart from analytical analysis, a rewarming test of Jurkat cell was performed to experimentally evaluate the electromagnetic rewarming technology. A comprehensive section on cryopreservation of large volume of stem cell has been tried to prepare, and it is aimed to provide insights about rapid-uniform rewarming during cryopreservation.

Author details

Shen Ren[1], Zhiquan Shu[1,2*], Jiaji Pan[3], Ji Peng[1], Junlan Wang[1], Chunhua Zhao[4] and Dayong Gao[1*]

1 Department of Mechanical Engineering, University of Washington, Seattle, WA, USA

2 School of Engineering and Technology, University of Washington Tacoma, Tacoma, WA, USA

3 School of Engineering and Design, Hunan Normal University, Changsha, Hunan, China

4 School of Life Sciences, Shanghai University and Institute of Basic Medical Sciences, Chinese Academy of Medical Sciences, China

*Address all correspondence to: zqshu@uw.edu and dayong@uw.edu

References

[1] Bruder, S.P., Jaiswal, N. and Haynesworth, S.E. (1997), Growth kinetics, self-renewal, and the osteogenic potential of purified human mesenchymal stem cells during extensive subcultivation and following cryopreservation. J. Cell. Biochem., 64: 278-294. doi:10.1002/(SICI)1097-4644(199702)64:2<278::AID-JCB11>3.0.CO;2-F

[2] Pegg DE. Principles of cryopreservation. Methods in molecular biology (Clifton, NJ). 2015;1257:3.

[3] Mehl PM. Nucleation and Crystal Growth in a Vitrification Solution Tested for Organ Cryopreservation by Vitrification. Cryobiology. 1993;30(5):509-518. doi:10.1006/cryo.1993.1051

[4] Mazur, P, Leibo, S.P, Chu, E.H.Y. A two-factor hypothesis of freezing injury: Evidence from Chinese hamster tissue-culture cells. Experimental Cell Research. 1972;71(2):345-355. doi:10.1016/0014-4827(72)90303-5

[5] Gao, D, Critser, J. K. Mechanisms of Cryoinjury in Living Cells. ILAR journal. 2000;41(4):187-196. doi:10.1093/ilar.41.4.187

[6] Mazur P, Mazur. Freezing of living cells: mechanisms and implications. American journal of physiology. 247(3 Pt 1):C125-C142. doi:10.1152/ajpcell.1984.247.3.C125

[7] Mazur P. Cryobiology: The Freezing of Biological Systems. Science (American Association for the Advancement of Science). 1970;168(3934):939-949. doi:10.1126/science.168.3934.939

[8] Bank H. Visualization of freezing damage. II. Structural alterations during warming. Cryobiology. 1973;10(2):157-170. doi:10.1016/0011-2240(73)90023-0

[9] Mackenzie, A.P. (2008). Death of Frozen Yeast in the Course of Slow warming. In Ciba Foundation Symposium - The Frozen Cell (eds G.E.W. Wolstenholme and M. O'Connor). doi:10.1002/9780470719732.ch6

[10] Fahy, Gregory M, Wowk, Brian, Wu, Jun. Cryopreservation of Complex Systems: The Missing Link in the Regenerative Medicine Supply Chain. Rejuvenation research. 2006;9(2):279-291. doi:10.1089/rej.2006.9.279

[11] Karlsson, Jens O.M, Toner, Mehmet. Long-term storage of tissues by cryopreservation: critical issues. Biomaterials. 1996;17(3):243-256. doi:10.1016/0142-9612(96)85562-1

[12] Brockbank KG, Chen Z, Greene ED, Campbell LH. Vitrification of heart valve tissues. Methods Mol Biol. 2015;1257:399-421. doi:10.1007/978-1-4939-2193-5_20

[13] Kuwayama M. Highly efficient vitrification for cryopreservation of human oocytes and embryos: The Cryotop method. Theriogenology. 2007;67(1):73-80. doi:10.1016/j.theriogenology.2006.09.014

[14] Ketterer, F.D, Holst, H.I, Lehr, H.B. Improved viability of kidneys with microwave thawing. Cryobiology. 1971;8(4):395-395. doi:10.1016/0011-2240(71)90197-0

[15] Guttman, Frank M, Lizin, Jacques, Robitaille, Pierre, Blanchard, Hervé, Turgeon-Knaack, Claire. Survival of canine kidneys after treatment with dimethyl-sulfoxide, freezing at −80 °C, and thawing by microwave illumination. Cryobiology. 1977;14(5):559-567. doi:10.1016/0011-2240(77)90166-3

[16] Pegg, David E, Green, Colin J, Walter, Clive A. Attempted canine renal cryopreservation using

dimethyl sulphoxide helium perfusion and microwave thawing. Cryobiology. 1978;15(6):618-626. doi:10.1016/0011-2240(78)90086-X

[17] Burdette, E.C, Karow, A.M, Jeske, A.H. Design, development, and performance of an electromagnetic illumination system for thawing cryopreserved kidneys of rabbits and dogs. Cryobiology. 1978;15(2):152-167. doi:10.1016/0011-2240(78)90020-2

[18] Ruggera, Paul S, Fahy, Gregory M. Rapid and uniform electromagnetic heating of aqueous cryoprotectant solutions from cryogenic temperatures. Cryobiology. 1990;27(5):465-478. doi:10.1016/0011-2240(90)90035-3

[19] Rachman, M.J, Evans, S, Pegg, D.E. Experimental results on the rewarming of a cryopreserved organ phantom in a UHF field. Journal of Biomedical Engineering. 1992;14(5):397-403. doi:10.1016/0141-5425(92)90085-Y

[20] Robinson, M.P, Pegg, D.E. Rapid electromagnetic warming of cells and tissues. IEEE Transactions on Biomedical Engineering. 1999;46(12):1413-1425. doi:10.1109/10.804569

[21] Robinson, Martin P, Wusteman, Monica C, Wang, Lihong, Pegg, David E. Electromagnetic re-warming of cryopreserved tissues: effect of choice of cryoprotectant and sample shape on uniformity of heating. Physics in Medicine and Biology. 2002;47(13):2311-2325. doi:10.1088/0031-9155/47/13/309

[22] Wusteman, Monica C, Pegg, David E, Robinson, Martin P, Wang, Li-Hong, Fitch, Paul. Vitrification media: toxicity, permeability, and dielectric properties. Cryobiology. 2002;44(1):24-37. doi:10.1016/S0011-2240(02)00002-0

[23] Eisenberg, David P, Taylor, Michael J, Rabin, Yoed. Thermal expansion of the cryoprotectant cocktail DP6 combined with synthetic ice modulators in presence and absence of biological tissues. Cryobiology. 2012;65(2):117-125. doi:10.1016/j.cryobiol.2012.04.011

[24] Pegg, David E, Wusteman, Monica C, Boylan, Serena. Fractures in Cryopreserved Elastic Arteries. Cryobiology. 1997;34(2):183-192. doi:10.1006/cryo.1996.1997

[25] Hasar, Ugur Cem, Oral, Emin Argun. A METRIC FUNCTION FOR FAST AND ACCURATE PERMITTIVITY DETERMINATION OF LOW-TO-HIGH-LOSS MATERIALS FROM REFLECTION MEASUREMENTS. Electromagnetic waves (Cambridge, Mass). 2010;107:397-412. doi:10.2528/PIER10071308

[26] Ansorge, Sven, Esteban, Geoffrey, Schmid, Georg. Multifrequency permittivity measurements enable on-line monitoring of changes in intracellular conductivity due to nutrient limitations during batch cultivations of CHO cells. Biotechnology Progress. 2010;26(1):272-283. doi:10.1002/btpr.347

[27] Juan-García, P, Torrents, J M. Measurement of mortar permittivity during setting using a coplanar waveguide. Measurement Science and Technology. 2010;21(4):045702. doi:10.1088/0957-0233/21/4/045702

[28] Lee, K. Y, Abbas, Z, Yeow, Y. K, Nur Sharizan, M. D, Meng, C. E. In situ measurements of complex permittivity and moisture content in oil palm fruits. European physical journal Applied physics. 2010;49(3):31201. doi:10.1051/epjap/2010007

[29] John H. Bradford, Joel T. Harper, Joel Brown. Complex dielectric permittivity measurements from ground-penetrating radar data to estimate snow liquid water content in the pendular regime. Water Resources

Research. 2009;45(8):W08403-n/a. doi:10.1029/2008WR007341

[30] Ba, Doudou, Sabouroux, Pierre. EpsiMu, A toolkit for permittivity and permeability measurement in microwave domain at real time of all materials: Applications to solid and semisolid materials. Microwave and Optical Technology Letters. 2010;52(12):2643-2648. doi:10.1002/mop.25570

[31] Zohdi, T.I, Kuypers, F.A, Lee, W.C. Estimation of red blood cell volume fraction from overall permittivity measurements. International Journal of Engineering Science. 2010;48(11):1681-1691. doi:10.1016/j.ijengsci.2010.04.013

[32] Hasar, Ugur Cem, Yurtcan, Mustafa Tolga. A microwave method based on amplitude-only reflection measurements for permittivity determination of low-loss materials. Measurement. 2010;43(9):1255-1265. doi:10.1016/j.measurement.2010.07.002

[33] Shibata K. Measurement of Complex Permittivity for Liquid Materials Using the Open-Ended Cut-Off Waveguide Reflection Method. IEICE transactions on electronics. 2010;E93-C(11):1621-1629. doi:10.1587/transele.E93.C.1621

[34] V. Komarov, Wang, S., Tang, J., Permittivity and measurements, in: K. Chang (Ed.) Encyclopedia of RF and microwave engineering (3693-3711), John Wiley and Sons, Inc, DOI: 10.1002/0471654507

[35] Mi Lin, Duane, M.H, Afsar, M.N. Cavity-Perturbation Measurement of Complex Permittivity and Permeability of Common Ferrimagnetics in Microwave-Frequency Range. IEEE Transactions on Magnetics. 2006;42(10):2885-2887. doi:10.1109/TMAG.2006.879885

[36] Waldron, R.A.: 'Perturbation theory of resonant cavities', Proceedings of the IEE - Part C: Monographs, 1960, 107, (12), p. 272-274, DOI: 10.1049/pi-c.1960.0041

[37] Robinson, M.P, Pegg, D.E. Rapid electromagnetic warming of cells and tissues. IEEE Transactions on Biomedical Engineering. 1999;46(12):1413-1425. doi:10.1109/10.804569

[38] Liang, Xin M, Sekar, Praveen K, Zhao, Gang, et al. High accuracy thermal conductivity measurement of aqueous cryoprotective agents and semi-rigid biological tissues using a microfabricated thermal sensor. Scientific reports. 2015;5(1):10377. doi:10.1038/srep10377

[39] Evans, S, Rachman, M.J, Pegg, D.E. Design of a UHF applicator for rewarming of cryopreserved biomaterials. IEEE Transactions on Biomedical Engineering. 1992;39(3):217-225. doi:10.1109/10.125006

[40] Luo, Dawei, Yu, Chun, He, Liqun, Lu, Caicheng, Gao, Dayong. Development of a single mode electromagnetic resonant cavity for rewarming of cryopreserved biomaterials. Cryobiology. 2006;53(2):288-293. doi:10.1016/j.cryobiol.2006.07.001

[41] Bai, X, Pegg, D.E, Evans, S, Penfold, J.D.J. Analysis of electromagnetic heating patterns inside a cryopreserved organ. Journal of Biomedical Engineering. 1992;14(6):459-466. doi:10.1016/0141-5425(92)90097-5

[42] Cai-Cheng Lu, Huai-Zhi Li, Dayong Gao. Combined electromagnetic and heat-conduction analysis of rapid rewarming of cryopreserved tissues. IEEE Transactions on Microwave Theory and Techniques. 2000;48(11):2185-2190. doi:10.1109/22.884213

[43] Penfold, J.D.J, Evans, S. Control of Thermal Runaway and Uniformity of Heating in the Electromagnetic Rewarming of a Cryopreserved Kidney Phantom. Cryobiology.

1993;30(5):493-508. doi:10.1006/cryo.1993.1050

[44] Han, X, Gao, D. Y, Luo, D, Yu, C, Lu, C. C. Numerical simulation of the microwave rewarming process of cryopreserved organs. Microwave and Optical Technology Letters. 2005;46(3):201-205. doi:10.1002/mop.20945

[45] Vaz, Raquel H, Pereira, José M.C, Ervilha, Ana R, Pereira, José C.F. Simulation and uncertainty quantification in high temperature microwave heating. Applied Thermal Engineering. 2014;70(1):1025-1039. doi:10.1016/j.applthermaleng.2014.06.005

[46] Basak, Tanmay, Aparna, K, Meenakshi, A, Balakrishnan, A.R. Effect of ceramic supports on microwave processing of porous food samples. International Journal of Heat and Mass Transfer. 2006;49(23):4325-4339. doi:10.1016/j.ijheatmasstransfer.2006.05.012

[47] Zhang, H, Datta, A. K, Taub, I. A, Doona, C. Electromagnetics, heat transfer, and thermokinetics in microwave sterilization. AIChE Journal. 2001;47(9):1957-1968. doi:10.1002/aic.690470907

Chapter 3

Umbilical Cord Blood and Cord Tissue Bank as a Source for Allogeneic Use

Tokiko Nagamura-Inoue and Fumitaka Nagamura

Abstract

Recently, umbilical cord blood (CB) has received attention as the allogeneic optimum source for immunotherapies. More recently, the umbilical cord (UC) has been rapidly utilized as an abundant source of mesenchymal stromal cells (MSCs), which migrate toward the inflammatory and damaged tissue to subside the inflammation and support tissue repair. Both CB and UC can be provided "off-the-shelf" cell products for immunotherapies and regenerative medicine. As biomedical wastes, CB and UC can be obtained noninvasively without any risks to the donor. CB cells and UC-derived MSCs (UC-MSCs) also have higher proliferation potentials than other cells obtained from adult tissues. In addition, UC-MSCs are less immunogenic and have significant immunosuppressive ability. Several clinical trials with CB or UC-MSCs have been conducted based on these advantages. The establishment of a stable supply system of CB and UC-MSCs is critical now for their utilization in regenerative and immune cell therapies. We have thus established the cord blood/cord bank, "IMSUT CORD," as a new type of biobank, to supply both frozen CB and UC tissues and derived cells for research and clinical uses. In this chapter, we will introduce the overall flow from collection to shipment and discuss several issues that need to be resolved in unrelated allogeneic stable supply system.

Keywords: cord blood, umbilical cord, mesenchymal stromal cells, immune cell therapy, regenerative medicine, biobank, explant method, large-scale expansion

1. Introduction

Umbilical cord blood (CB) has been utilized as a source of hematopoietic stem cells for three decades. It could potentially also serve as the optimum source of immune cells, such as mononuclear cells (MNC), regulatory T cells, NK cells, and mesenchymal stromal cells (MSCs) with or without genetic modifications, for immunotherapy and neurogenic regeneration in some cases. In addition, UB could be prepared as readily available products [1].

It is well-known that human mesenchymal stromal cells (MSCs) can be harvested from various tissues that include the bone marrow (BM) [2], cord blood (CB) [3], adipose tissue [4], placenta [5], and umbilical cord (UC) [6]. Recently, clinical trials using MSCs for various diseases have been conducted, and some of them were approved. The BM is considered the traditional source of MSCs, and the characteristics and applications of BM-derived MSCs (BM-MSCs) have been studied in detail. However, the harvesting of these cells is associated with an invasive procedure, and it may cause hemorrhage, infection, and chronic pain. In addition, BM-MSCs exhibit accelerated senescence as the donors' age [7].

On the other hand, both CB and UC are routinely discarded as medical waste. The harvesting of CB and UC-derived MSCs (UC-MSCs) is therefore noninvasive and painless. The CB drawn from the UC and placenta is well-known as the source of hematopoietic stem cells for CB transplantations. However, in this article, we focus on the CB as the source for immune cells and regenerative medicine, such as regulatory T cells (Treg), NK cells, MSCs, and so on. The UC is the conduit between the developing embryo and placenta and consists of two umbilical arteries and one umbilical vein buried in the Wharton's jelly. UC-MSCs have the abilities of multipotency and self-renewal properties comparable or superior to MSCs derived from other tissues in some papers. For this reason, several private CB banks have begun to collect CB and UC. We have thus established the cord blood/cord bank, "IMSUT CORD", as a new type of biobank, to supply both frozen UC tissues and master cells for research and clinical uses.

In this chapter, we will introduce the overall flow from collection to shipment as taking the example of IMSUT CORD and discuss several issues that need to be resolved in unrelated allogeneic off-the-shelf stable supply system at present.

2. Cord blood and umbilical cord collection

There are many public and private CB banks in the world, in which procedures are nearly standardized intended for hematopoietic stem cell transplantation (HSCT) as shown in Section 5. The procedures include informed consent acquisition from the mother, collection of CB, processing, storage, to shipment, which have been already established. In the case of UC bank for unrelated allogeneic uses, the International Council for Harmonization of Technical Requirements for Pharmaceuticals for Human Use (ICH) which issued ICH Q5A as the regulation of materials for biological products requires the second blood test from the baby's mother, to deny viral infection in window period at delivery. **Figure 1** shows the overall process of banking in the mother's side. We deal with both CB and UC.

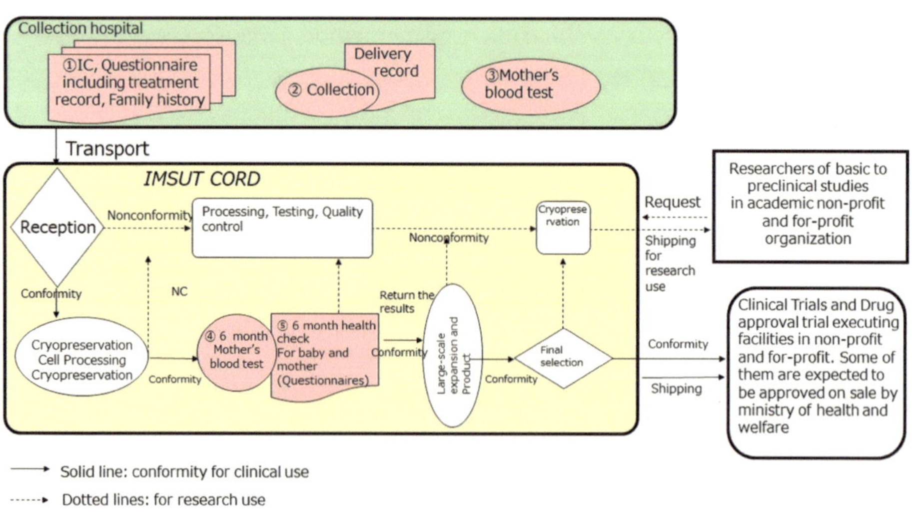

Figure 1.
Overall flow from informed consent acquisition to shipment.

In the CB and UC collection hospitals, the purpose, overall process, private information policy, the right to withdrawal, 6-month health check, and second blood test for the mother to confirm the negative study of infection are explained to the mother, and she gives written consent as the guardian of the baby. In addition to obtaining informed consent, questionnaires about medical history, genetic history of the baby donor's family, and history for the mother's communicable disease risk behavior are conducted to survey their health. The CB and UC are then collected, and the mother's blood is tested for infections. These documentations and tests in CB bank can be referred to UC banking as well, although additional safety tests for UC banking shall be required strictly. UC-MSCs from one donor can be delivered and administered to many patients. Especially when the CB and UC passed the safety and some quality tests at clinical grade, the mother is asked to receive the blood test to make sure that all infection tests are negative in 6 months after delivery. These second tests are demanded by the Pharmaceuticals and Medical Devised Agency (PMDA) like the FDA and EMA, because it should be proven that the donor's mother and baby were not in the window period of viral infections at delivery. On the other hand, bacterial contamination is also taken care because the baby and placenta with UC come out from nonsterile vagina. We collect UC in the case of a scheduled cesarean section to reduce the possibility of contamination due to the exposure to the vaginal bacterial/fungal flora.

CB and UC are then transported from the collection hospitals to the CB/UC bank under controlled and validated temperature. CB is transported at validated temperature range (2–25°C) to protect cell viability, and the UC is cooled at 2–10°C in our facility.

3. Cord blood processing

Among the processing methods to obtain nucleated cells from CB for hematopoietic stem cell transplantation, the hydroxyethyl starch (HES) centrifugation method (HES method) is the most efficient and common. The HES method originated from the New York Blood Center CB bank [8]. Recently, automated CB processing SEPAX® (GE Healthcare Life Sciences) [9] and AutoXpress Platform® (Cesca Therapeutics, Inc.) [10] have been developed. For CB cryopreservation, DMSO and Dextran 40 together with CB-plasma are used worldwide [8].

On the other hand, no processing method of fresh or frozen CB not for hematopoietic stem cell transplantation has been settled as standard. The use of mononuclear cells (MNCs) obtained by the Ficoll-Paque centrifugation method (Ficoll method) or cell sorting with antibody-conjugated magnetic beads might be a new candidate for further processing and culture method. CB processed by HES method resulted in whole nucleated cells including neutrophils, monocytes, lymphocytes, and nucleated red blood cells with some amount of red blood cells (RBC). The recovery rate of hematopoietic stem cells and mononuclear cells processed by HES method is superior to those by Ficoll method. That is why HES method is introduced by CB banks in the world [8]. However, neutrophils in the nucleated cells and residual RBC may cause the aggregation resulting in the difficulty of further processing, when the frozen and thawed cells are diluted with large volume of isotonic solutions such as medium. Only frozen-thawed CB nucleated cells can be diluted with dextran and albumin/saline solution [8]. On the other hand, frozen-thawed MNCs processed by Ficoll method does not require such a special solution and can be diluted with medium and PBS, although the recovery rate of MNCs from the fresh CB by Ficoll method is less than that by HES methods.

MSCs derived from fresh CB are difficult to expand. Only one company, Medipost Co., Ltd., in Korea, has succeeded in expanding CB-derived MSCs. Their product, Cartistem®, has been approved by the Ministry of Food and Drug Safety in Korea for the treatment of osteoarthritis [11].

4. Umbilical cord processing

There are diverse procedures and culture methods for the isolation of MSCs from the various compartments of UC, such as Wharton's jelly, veins, arteries, UC lining membrane, subamnion, and perivascular regions [12]. The isolation methods of MSCs from the Wharton's jelly, vein, and arteries of UC are reported previously, although the marked differences were not found as far as the 10% fetal bovine serum (FBS) and α minimum essential medium (MEM) [13]. There are several papers to obtain MSCs from whole UC versus Wharton's jelly [14] or different parts of the same UC [15], but we suggest that to process from whole UC seems sufficient and simple for further processing [15]. Despite the wide range of isolation and culture procedures, the different groups seem to agree on the cryopreservation of UC tissue [16] and explant method [17] followed by the harvest of migrating cells from tissue. However, large-scale culture methods remain to be determined. **Figure 2** shows the example of scheme of CB and UC collection and process and shipping to clinical use.

4.1 Cryopreservation of UC tissue

It is known that the UC tissue can be frozen in a cryoprotectant. This possibility of cryopreservation is the advantages of UC tissue for both clinical and research uses. The reasons of the advantages are:

1. UC tissue processing can be started after the donor's health and infection statuses are confirmed well. This leads to initial cost-effectiveness because unnecessary works using inappropriate materials are eliminated. In addition, we can thaw a small amount of the UC to survey, before culturing MSCs in a large scale.

2. Storage of the tissues of origin allows us to keep traceability and to check the quality as the biological resources at a later date.

3. When new reagents or techniques were developed in the future, we can isolate novel cells from the cryopreserved UC tissues.

4. If the donor, the baby, has diseases that can be treated with autologous cells, including iPS cells or gene-modified cells, or autologous UC-MSCs, the cryopreserved UC tissues would be the appropriate source.

Several animal serum-free cryoprotectants containing 5–10% dimethyl sulfoxide (DMSO) are available. Whether the use of serum originating from animals, such as fetal bovine serum (FBS), is required, is critical, because it adds the risk of the transmission of zoonotic infections, immunological reactions, and additional regulatory issues [18]. There are several reports of cryopreservation of the UC tissue, using serum-free and xenogeneic animal-free (xeno-free) cryoprotectants. Ennis et al. introduced CryoStor CS10® (BioLife Solutions Inc., WA) for isolating human UC perivascular cells (HUPVCs). However, they did not show the comparative test results to those of fresh UC [19]. Roy et al. reported the cryopreservation of the

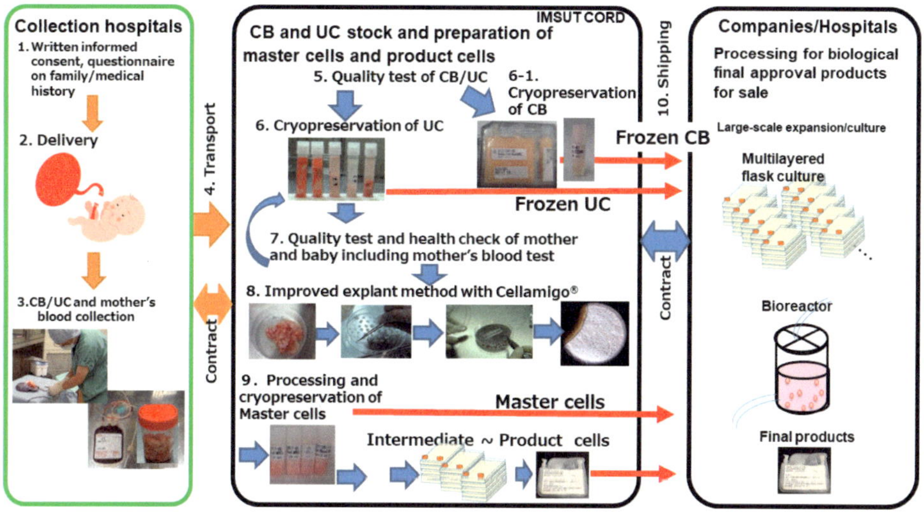

Figure 2.
IMSUT CORD scheme of CB and UC from collection to shipping for clinical use.

UC tissue in 10% DMSO and 0.2 M sucrose solution, but the cumulative cell yield derived from the frozen-thawed UC-MSCs in their solution was inferior to that of fresh UC-MSCs [20]. We previously reported the cryopreservation of UC tissue, with no impact on viability, using a serum- and animal origin-free cryoprotectant, STEM-CELLBANKER® [16]. We demonstrated that cells derived from UC cryopreserved in this manner retained the phenotypic characteristics of MSCs, including the immunosuppressive activity in allogeneic mixed lymphocyte reactions, as well as differentiation potential. As shown in **Figure 2**, with the cryopreservation of UC tissue, UC processing might be altered.

4.2 Improved explant method

There are two major approaches after frozen-thawed UC tissue: explant and enzymatic digestion methods. Frozen-thawed UC tissue is manually minced into small fragments approximately 1–2 mm^3 in size. Mincing is preferred to using a surgical scalpel or the use of an autologous mixer. These fragments are aligned and seeded regularly on a tissue culture-treated dish. After the tissue fragments attach to the bottom of the dish, culture media is added, slowly and gently in order not to detach the fragments [21–24]. Media is then refreshed every 3–7 days for 2–4 weeks until the fibroblast-like adherent cells reach 80–90% confluence. Subsequently, adherent cells and tissue fragments are rinsed once with PBS, detached using trypsin, and washed with media. The culture is then filtered to remove tissue fragments. The disadvantage in the explant method is that tissue fragments often float in media, resulting in the poor recovery of cells. To protect the exfoliation of tissue fragments from the bottom of the culture dish, we introduced stainless steel mesh (Cellamigo®; Tsubakimoto Chain Co.) shown in **Figure 2**, No. 8. In this manner, we can plate source tissue more quickly and harvest more MSCs. In addition, the incubation time required to reach 80–90% confluence is reduced [17].

In the enzymatic digestion method, UC is minced into small pieces and immersed in the media containing enzymes such as collagenase, or a combination of collagenase and hyaluronidase with or without trypsin [21, 24–26]. The cells dissociated by the enzymes are then cultured until they reach full confluence.

However, the digestion method is costly and time-consuming and may result in decreased cell viability due to lytic activity and varying sensitivity of the cells to collagenase. In addition, the initial harvested cells include more of the other types of cells compared with that harvested using the explant method.

4.3 Large-scale expansion and harvesting the cells

It is critical to consider how much we can expand the UC-MSCs to allow allogeneic "off-the-shelf" industrial availability, because the proliferation of adherent cells needs a large surface area. The conventional method uses multi-layered flasks, and the cells are cultured in incubators installed in cleanrooms. These multilayer flasks can consistently support the expansion of UC-MSCs, and the state of cell confluence can be examined under the microscope. However, this method requires the considerable involvement of operators because the processes of seeding, refreshment, passage, and harvest require individual and manual works. Several companies have introduced the spinner bioreactor with a microcarrier made of plastic, dextran, denatured collagen-coated beads, and other components. The bioreactor system may reduce the number of operators required and may allow to reduce the clean levels of facility since it is a closed system. On the contrary, several critical problems of the bioreactor system exist. The cost of equipment is high and it is difficult to evaluate cell proliferation environment such as pH, lactate, and so on. When some microcarriers are torn off by spinner, or undigested microcarriers are residual in the final products, we have no ideas to remove the residual microcarriers completely. Recently, a plastic bag bioreactor system with a microcarrier in gentle locking was reported [27]. The most critical problem is that the cells produced by the flask-based culture method may be different from those by bioreactors. Harvesting cells on a large scale is still not easy. Recently, filter-based cell concentration and washing systems were introduced (https://www.kaneka.co.jp/en/business/healthcare/med_006.html, KANEKA, Japan). Automatic cell packaging may be also required in large-scale expansion.

The academic culture level such as IMSUT CORD is at small to middle scale. Only the company may have the ability to expand the cells at extra-large scale and maintain to control and supply the cell product for clinic constantly.

4.4 Long-term cryopreservation

Master and product cells of UC-MSCs for clinical use are usually required for long-term cryopreservation, together with records on the donor infant and the mother. There are several cryoprotectants for long-term cryopreservation. The most popular cryoprotectant consists of 5–10% dimethyl sulfoxide with human albumin. Recently, serum-free cryoprotectants, described in Section 4.1, have been commercialized and are thought to be more ideal compared to those containing human-derived serum. In addition to cryoprotectants, it is important to build an adequate record preservation system. Those who manage the long-term cryopreservation should preserve the records that include the documentation relating to the collection including donors, processing, results of quality tests, and instruments directly related to the products. The kinds of records and the length to be preserved are in accordance with the bank policies and standards and the corresponding domestic laws and regulations. It is necessary to discuss how long we should or we can cryopreserve CB and UC tissue, UC master cells, and product cells, in the technical and ethical aspects. In the technical aspect, the cell-preserving vessel to accommodate the cell suspension for long-term freezing should be durable in a liquid nitrogen. In the ethical aspect, we do not expect whether the babies can recapture their ownership of CB and UC even though their mother waived the ownership of them, when they grow up to be adult.

4.5 Quality and safety assurance of UC-MSCs

The Mesenchymal and Tissue Stem Cell Committee of the International Society for Cellular Therapy proposed the minimal criteria for defining human MSCs [28, 29]. First, MSCs must be plastic-adherent when maintained in standard culture conditions. Second, MSC must express CD105, CD73, and CD90 but not CD45, CD34, CD14 or CD11b, CD79α or CD19, and HLA-DR surface molecules. Third, MSCs must differentiate into adipocytes, chondroblasts, and osteoblasts in vitro [30–32]. Immunosuppressive effects have now become the most popular property of MSCs for potential clinical use [12]. Defect in HLA-class II expression with negative CD80 and CD86in UC-MSCs even in the presence of inflammatory cytokines such as IFN-γ can theoretically rescue them from immune recognition by CD4+ T cells [33]. MSCs can also inhibit proliferation of and cytokine secretion by immune cells, as well as alter subtypes of macrophage from M1 to M2 in vitro [34–37]. This immunomodulation is linked mainly to soluble factors such as indoleamine 2,3-dioxygenase (IDO), PGE2, and HLA-G5 [38], hepatocyte growth factor, and transforming growth factor-β1 released from MSCs [39]. Further quality tests are dependent on clinical applications and characteristics of MSCs.

The safety tests required differ according to the risks of clinical applications. For example, the tests of CB banking for hematopoietic stem cell transplantation are different from those of UC-MSCs. Donor-recipient relation of the former is one-to-one, and the risk is limited to one patient. On the other hand, that relationship of the latter, as UC-MSCs master cells and product cells, is one-to-many, so hundreds of patients may suffer health injuries by one donor. Thus, the vials of UC-MSCs are tested thoroughly at a designated time not only known viruses but also unknown viruses. Those tests should follow the local, national or international applicable laws and regulations. More precise safety tests for CB and UC shall be described elsewhere for the respective products for clinical application.

5. Standards and guidance for CB and UC from collection to release

There are international standards/guidance for CB collection, banking, and release of hematopoietic stem cell transplantation, such as the Foundation for the Accreditation of Cellular Therapy (FACT)/NETCORD [40], American association of Blood Banks (AABB), US Food and Drug Administration (FDA) shown in **Table 1**, and local standards or regulations under the applicable laws in respective countries. A CB/UC bank, facility, or individual should implement if the standard of practice in the community or applicable law establish additional requirements. International standards/guidance for biobanking process for UC collection, processing, culture, and release has not been settled, but collection and banking protocols can follow the CB banking standards and good tissue practice. Each CB/UC bank, facility, and individual should analyze its practices and procedures to determine whether additional standards apply. Compliance with the standards is not an exclusive means of complying with the standard of care in the industry or community or with local, national, or international laws or regulations [40]. Allogeneic public CB banks requested US FDA accreditation with FACT/NETCORD or AABB in the USA, while the CB banks in Europe (EU) required FACT/NetCord

Items	Accreditation organization	Standards or guidance titles
Cord blood (CB) processing for hematopoietic stem cell transplantation	FACT/NETCORD FACT/JACIE	International Standards for Cord Blood Collection, Banking, and Release for Administration International Standards for Hematopoietic Cellular Therapy Product Collection, Processing, and Administration http://www.factwebsite.org/cbstandards/
	FDA in the USA	Guidance for Industry: Minimally Manipulated, Unrelated Allogeneic Placental/Umbilical Cord Blood Intended for Hematopoietic Reconstitution for Specified Indications Guidance for Industry and FDA Staff: Investigational New Drug Applications for Minimally Manipulated, Unrelated Allogeneic Placental/Umbilical Cord Blood Intended for Hematopoietic and Immunologic Reconstitution in Patients with Disorders Affecting the Hematopoietic System http://www.fda.gov/BiologicsBloodVaccines/GuidanceComplianceRegulatoryInformation/Guidances/default.htm.
	AABB	Standards for Cellular Therapy Services http://www.aabb.org/sa/Pages/Standards-Portal.aspx
Umbilical cord-derived cells including mesenchymal stromal cells (UC-MSCs)/somatic cell or other derivative cells CB not intended for hematopoietic stem cell transplantation	FACT	Common standards for Cellular Therapies http://www.factwebsite.org/cbstandards/
	FDA in the USA	Good Tissue Practice 21CFR 1271.210 Current Good Tissue Practice (CGTP) and Additional Requirements for Manufacturers of Human Cells, Tissues, and Cellular and Tissue-Based Products (HCT/Ps) Guidance for Industry: Preclinical Assessment of Investigational Cellular and Gene Therapy Products Guidance for FDA Reviewers and Sponsors Content and Review of Chemistry, Manufacturing, and Control (CMC) Information for Human Somatic Cell Therapy Investigational New Drug Applications (INDs) http://www.fda.gov/cber/guidelines.htm
	AABB	Standards for Cellular Therapy Services http://www.aabb.org/sa/Pages/Standards-Portal.aspx
	EMA in EU	Tissues and Cells Directives: Guideline on human cell-based medicinal products (EMEA/CHMP/410896/2006) for ATMP
	PMDA (Japan)	Good Gene, Cellular, and Tissue-based Products Manufacturing Practice (GCTP)
Quality management system	ISO	ISO9001 ISO/TC276 (Draft) https://www.iso.org/standard/62085.html

Foundation for the Accreditation of Cellular Therapy, FACT; US Food and Drug Administration, FDA; American Association of Blood Banks, AABB; advanced therapy medicinal products, Pharmaceutical and medical devices agency (PMDA). This table does not include the law defined in each country. These standards, guidance, guidelines, and practices are not intended to apply all cell therapies using CB and UC. The CB/UC bank carefully chooses and implements them for your intended products under the applicable law.

Table 1.
Standards or guidance related to cord blood and umbilical cord-derived cells.

with additional requirements like FACT/JACIE standards, when it is requested by the respective national regulation affairs. There are many private or private-public combined CB banks in the world, which tend to follow the AABB standards and have the inspection and accreditation (http://www.aabb.org/sa/facilities/celltherapy/Documents/AABB-Accredited-Cord-Blood-Facilities.pdf).

6. Special issues of CB and UC bank system for allogeneic use

The number of clinical trials using CB and UC-MSCs in the fields of immune cell therapies and regenerative medicine has been increasing. On the other hand, CB as a source of hematopoietic stem cell transplantation is less used recently, because the cell number is limited, the engraftment of HSC is delayed, and HLA haplo-identical HSCT is induced and controlled. These clinical trials are aimed uses that include severe acute graft-versus-host disease (GVHD) treatment, rapid engraftment of HSCT, and the prevention of severe acute GVHD. Clinical trials using CB- and UC-MSCs are summarized in **Tables 2** and **3**, respectively. We started a sponsor-investigator clinical trial using UC-derived MSCs for patients with treatment-resistant severe acute GVHD supported by the research fund of the Japan Agency for Medical Research and Development (AMED). Consistent supply is the critical key to conduct clinical trials and for marketing. For the stable supply of frozen CB and UC, or UC-derived MSCs, we have established a CB and UC bank, named IMSUT CORD, in our institute. This bank also provides CB and UC-MSCs for immunotherapy and regenerative medicine products to hospitals and pharmaceutical companies shown in **Figure 2**. The bank also provides frozen CB, frozen UC, master cells, and the cells after master cells as an intermediate products requiring further processing or more culture in the companies.

The following are also the major points for managing CB and UC banking.

First, to build an adequate quality management system to serve the resource of cell therapy products, we have introduced the concept of the ISO 9001 and obtained its certification, and as a result, we introduced the concept of PDCA cycle. Second, involvements of various kinds of specialists must be considered. There are many procedures, such as collection, obtaining informed consent, application to ethics review committee, and document management. Third, health check and infection test of the donor's mother are required to ensure that no infection is detected after window period of infections. In this process, both traceability and personal information protection must be satisfied. Fourth, we respect the right of decision to donate, rejection, or withdrawal. Donor's mother should be explained the policy of the bank that the consented withdrawal time is set at the initiation of processing for clinical use. Although the CB and UC belong to the baby, we obtain informed consent from the donor's mother as guardianship and ownership are asked to be transferred to the bank. Fifth, there is also the issue of how long the UC tissue and UC-MSCs can be cryopreserved. For example, in the Japanese public CB bank for HSCT, the CB is cryopreserved for 10 years as a clinical grade of HSC source. After this period, they are used for basic research or preclinical studies or discarded if they are not used for research. A cryopreservation period of 10 years for UC and UC-MSCs may be the first threshold to be checked. In addition, we disclose the information in website for the mothers who have not been explained about the new researches or new clinical trials at the first IC acquisition. Lastly, because unlike CB, the UC is a tissue considered as a part of the perinatal appendage, we must follow the tissue transport and medical disposal/waste regulations under the applicable laws or local rules and ethical standards.

Authors	Cell type	Disease	Patients number	Age (range) year	Route and procedure of administration	Cell number/kg or body	Results	Adverse events
Brunstein et al. [41]	CB-NC-derived Treg (CB from The New York Blood Center)	Grade II–IV acute GVHD	Treg: 11	61 (45–68)	IV	3–100 × 10^6 Treg/kg	aGVHD: Treg group 9%, control 45% cGVHD: Treg 0%, control 14%	No dose-limiting infusion adverse events
			Control: 22	60 (34–69)		—		
Kellner et al. [42]	Fucosylated or non-fucosylated UCB-Tregs	HSCT	5		IV (−1 day of HSCT)	1 × 10^6/kg		No infusion reactions
Zhu et al. [43]	CB-MNC	Chronic complete spinal cord injury	8 in Hong Kong	42.6 ± 2.7	IT (dorsal entry zone)	1.6–3.2 × 10^6	Walk 10 m, 15/20 pts. (p = 0.001), no necessity of assistance for bladder management, 12/20 (p = 0.001) and bowel management (p = 0.002)	1 neuropathic pain;1 subarachnoid hematoma and pneumocephalus due to cerebrospinal loss;1 arachnoid hemorrhage I HK group,
		Phase I–II	20 in Kunming	36.9 ± 2.4				68 AEs including postoperative wound swelling; 9 pain Overall 5 severe AE in 28 patients
Shah et al. [44]	CB-MNC-derived NK cells (CB from MD Anderson Cord Blood Bank)	Multiple myeloma undergoing autologous PBSCT	12	48–70		5 × 10^6, 1 × 10^7, 5 × 10^7 and 1 × 10^8 CB-NK cells/kg	10 achieved VGPR (8 near CR) as the best response	No infusional toxicities and no GVHD
Lv et al. [45]	CB-MSC + UC-MSCs	Autism	14 CB-MNC	CB-MNC: 7.08 (3.29–12.01)	IV	Proximately 2 × 10^6/kg CB-MNCs 1 × 10^6/kg of UC-MSCs 4 times in 5–7 day	Improvement of CARS, ABC scores, and CGI evaluation at 24 weeks in CB-MNCs with UC-MSCs	No treatment-related and no severe adverse effects
			9 CB-MNC and UC-MSCs	CB-MNC + UC-MSCs: 6.51 (3.98–9.83)				
			14 no cells therapy	Control: 5.02 (3.51–10.02)				

Authors	Cell type	Disease	Patients number	Age (range) year	Route and procedure of administration	Cell number/kg or body	Results	Adverse events
Dolstra et al. [46]	CB-CD34-derived NK cells (CB from Cord Blood Bank Nijmegen)	AML in old patients	10	68–76	IV	3 and 30 × 10^6/kg	NK cell maturation in vivo, MRD become negative in 2/4 with MRD before IV	No GVHD, no toxicity
Park et al. [47]	CB-derived MSCs	Rheumatoid arthritis	9	57.4 ± 10.0	IV	2.5 × 10^7, 5 × 10^7, or 1 × 10^8	DAS2/-ESR decreased, inflammatory cytokine levels are reduced	No DLT, no major toxicity
Laskowitz et al. [48]	CB-NC (CB from Carolinas Cord Blood Bank or MD Anderson Cord Blood Bank)	Cerebral stroke	10	65.5 (45–79)	IV on 3–9 days poststroke	Cell dose 1.54 (0.84–3.34) × 10^7/kg, $CD34^+$ 2.03 (0.10–6.80) × 10^5/kg	All improved by at least one grade in Modified Rankin Score	AE tolerated no serious AE
Huang et al. [49]	CB-MSCs	Cerebral palsy (age: 3–12)	27 (CB-MSCs)	CB-MSCs: 7 (3–12)	IV	27:4 CB-MSCs IV at 5 × 10^7 with basic rehabilitation treatment	Significant improved of GMFM-88 evaluation	No serious AE
			27 (control)	Control: 7 (3–12)				
Kim et al. [50]	CB-MSCs	Moderate-to-severe atopic dermatitis	34 (7 in phase I, 27 in phase IIa)	29.07 ± 2.03 (n = 14)	SC	2.5 × 10^7	Improved atopic dermatitis scores, pruritus score, serum IgE and eosinophil number	No serious AE
				28.08 ± 1.07 (n = 11)		5.0 × 10^7		

AE, adverse event; AML, acute myeloid leukemia; CB, cord blood; UC, umbilical cord; MSCs, mesenchymal stromal cells; MNCs, mononuclear cells; NK cells, natural killer cells; Treg, regulatory T cells; GVHD, graft-versus-host disease; PBSCT, peripheral blood stem cell transplantation; IV, intravenous injection; SC, subcutaneous injection.

Table 2.
Clinical trials using allogeneic cord blood.

Authors	Disease	Patients number	Age (range) year	Route and procedure of administration	Cell number/kg or body	Frequency interval	Results	Adverse events
Engraftment facilitation and graft-versus-host disease (GVHD) in hematopoietic stem cell transplantation								
Wu et al. [51]	Severe steroid-resistant aGVHD	2	Pt 1:4	IV	Pt 1: 3.3, 7.2, 8.0 × 10^6/kg	3	Improved	No
			Pt 2:6	IV	Pt 2: 4.1 × 10^6/kg	1		
Si et al. [52]	Severe aplastic anemia	37	5	IV (7–10 days after HSCT)	1 × 10^6/kg	1	aGVHD II–IV; 17 of 37 (45.9%)	No
			(0.75–11.58)				cGVHD, 7 of 37 (18.9%)	
Wu et al. [53]	Refractory/ relapsed hematologic malignancy	50	26 (9–58)	IV (4 h before haploidentical HSCT)	5 × 10^5/kg	1	aGVHD II–IV, 12 of 50 (24.0%) cGVHD, 17 of 45 (37.7%) (3 extended)	No
Wu et al. [54]	Severe AA	21	18 (4–31)	IV (4 h before HSCT)	5 × 10^5/kg	1	aGVHD II-IV;12 of 21 (57.1%) 3 of 9 extended cGVHD	No
Fu et al. [55]	Refractory severe AA	5	15.2 (9–22)	IV (2 days after PBSCT)	1 × 10^6/kg	1	No severe aGVHD or cGVHD	No
Gao et al. [56]	Prophylaxis of chronic GVHD after HLA-haploidentical stem cell transplantation	62	Age < 8, 15 pts.; 18–40, 39; >40, 8	IV	3 × 10^7 cells	Until cGVHD occurred, leukemia relapsed, or 4 cycles	cGVHD at 2 yr.: MSCs group 27.4%, control 49.0% (P = .021). Severe lung cGVHD: MSCs group 0, control 7 (P = .047)	No
Zhu et al. [57]	High-risk acute leukemia	25	11.2 (4–17)	IV (before haploidentical HSCT)	Median 1.14 × 10^6/kg (1.03–1.39 × 10^6/kg)	4 (over 7 days intervals)	aGVHDI, 8 of 25 (32.0%) cytomegalovirus viremia, 23 of 25 (92.0%)	No

Authors	Disease	Patients number	Age (range) year	Route and procedure of administration	Cell number/kg or body	Frequency interval	Results	Adverse events
Pan et al. [58]	Extensive bone marrow necrosis of a chronic myeloid leukemia patient	1	10	iBM IV	iBM: 2×10^7/kg IV: 2 pp. $\times 10^6$/kg	1	BM recovered	No
Neurogenic injuries								
Wang et al. [59]	Traumatic brain injury	20	27.5 (5–48)	Intrathecal (IT)	1×10^7	4 (5–7 days intervals)	Motor functional recovery after 6 months	No
Jin et al. [60]	Hereditary spinocerebellar ataxia	16	39.9 (21–56)	IV – intrathecal	IV; 4×10^7 IT; 2×10^7 cells	4 (over 7 days interval)	Motor functional recovery after 6 months	No
Wang et al. [61]	Cerebral palsy	16 (8 twins)	6.29 (3–12)	IT	1–1.5 $\times 10^7$ cells	4 (3–5 days intervals)	Motor functional recovery after 1 and 6 months	No
Diabetes mellitus								
Guan et al. [62]	DM (type 2)	6	40.5 (27–51)	IV	1×10^6/kg	2 (2 weeks interval)	Insulin-independent for 25–43 Mo, 3 dose reduction of insulin, others	No
Hu et al. [63]	DM (type 1)	15	17.6	IV	$2.6 \pm 1.2 \times 10^7$/kg	2 (4 weeks interval)	HbA1c and C-peptide improvement in MSCs group	No
Cai et al. [64]	DM (type 1)	21	18–29 (5–28) at onset	Supraselective pancreatic artery cannulation	1.1×10^6/kg, with autologous BM-MNC	1	Moderate improvement of metabolic measures	1 transient abdominal pain

Authors	Disease	Patients number	Age (range) year	Route and procedure of administration	Cell number/kg or body	Frequency interval	Results	Adverse events
Kong et al. [65]	DM (type 2)	18		IV	1×10^6/kg	Day 0 and until Day 90 if effective	FBS reduced plasma C-peptide and regulatory T cells increased	4/18: slight fever
Heart and angioplasty								
Cai et al. [66]	Avascular necrosis of the femoral head	30	41.6 (19–63)	Femoral head artery (co transplant with autologous BM)	Autologous BM-BM-MNCs, $60.7 \pm 11.5 \times 10^6$/kg UC-MSCs, 1×10^6/kg	1	Improved	No
Can et al. [67]	Myocardial ischemia	39	30–80	Intracoronary	2×10^7/kg	1	Ongoing	No
Zhao et al. [68]	Severe systolic heart failure	30	52.9 (20–79)	Intracoronary	Unknown	1	Cardiac remodeling and function improved with reduced mortality rate	No
Li et al. [69]	Coronary chronic total occlusion	15	Unknown	Intracoronary	$3 \times 10^6/4 \times 10^6/5 \times 10^6$/kg	1	Infarcted size reduced with improved left ventricular EF	No
Musialek et al. [70]	Acute myocardial infarction	10	55.6 (32–65)	Intracoronary	3×10^7/body	1	Feasible and procedural safe as off-the-shelf cellular therapy	Transient fever (38.9°C)

Authors	Disease	Patients number	Age (range) year	Route and procedure of administration	Cell number/kg or body	Frequency interval	Results	Adverse events
Bartolucci [71]	Heart failure	15	57.33 ± 10.05	IV	1×10^6 cells/kg	1	Significant improvements in LVEF, NYHA functional class, Minnesota Living with Heart Failure Questionnaire	No
Liver								
Xue et al. [72]	Decompensated liver cirrhosis	50	Unknown	Intrahepatic artery	3×10^7/body	1	Increase of serum albumin	No
Wang et al. [73]	Primary biliary cirrhosis	7	49 (33–58)	IV	5×10^5/kg at 4 weeks interval	3	ALP and γ-GTP	No
Shi et al. [74]	Prevention of acute liver allograft rejection	14 (13, single dose, 1 multiple dose)	57 ± 12	IV	1×10^6 cells	Single (13 pts), 3 times every 4w (1 pt)	Decreases of ALT, AST, T-BIL Histologic improvements, MSCs 6, control 0.	No
Liang et al. [75]	Liver cirrhosis caused by autoimmune diseases	23 (2 CB-MSC, 1 BM MSC)	53.4 (35–70)	IV	1×10^6 cells/kg	1	Not statistically significant improvement	2, fever; 3, mild fidgetiness, suffered from insomnia
Zhang et al. [76]	Ischemic-type biliary lesions following liver transplantation	12	47.3 ± 10.1	IV	1×10^6 cells/kg	6 (1, 2, 4, 8, 12, 16 weeks)	Significantly decreased need for interventional therapies. 1-, 2-yr graft survival rates: MSCs group (100%, 83.3%), control group (72.9%, 68.6%)	No

Authors	Disease	Patients number	Age (range) year	Route and procedure of administration	Cell number/kg or body	Frequency interval	Results	Adverse events
Xu et al. [77]	Hepatitis B virus-related acute-on-chronic liver failure	30:UC-MSC	UC-MSC: 40.67 ± 9.89	IV	10^5 cells/kg	UC-MSC, once a week, 4 times	No significant improvement of short-term prognosis	Fever, UC-MSC 11 pts., PE + UC-MSC 6 pts
		20, UC-MSC + plasma exchange	UC-MSC/ plasma exchange, 42.00 ± 6.55	IV		UC-MSC/ PE: first 2 UC-MSC: 2nd day after 1st, 3rd PE treatments		
Gastrointestinal tract								
Zhang et al. [78]	Crohn's disease	41	32.7 (20–41)	IV	1×10^6 cells/kg	Once a week, four times	Decreases of CDAI, HBI, corticosteroid dosage	Fever 4, upper respiratory tract infection, 7
Hu et al. [79]	Ulcerative colitis	34	42.9 ± 23.1	IV then IA	0.5×10^6 cells/kg	2, 7 days	Decreases of median Mayo score, histology score. Improvement of IBDQ scores	No
Skin								
Hashemi et al. [80]	Chronic skin ulcer	5	30–60	Covered by acellular amniotic membrane seeded with WJSCs	About 2×10^6 cells were seeded	Epithelial surface of acellular amniotic membrane	Significantly decreased wound healing time, wound size. Significantly declined wound size after 6, 9 days	Not stated

Authors	Disease	Patients number	Age (range) year	Route and procedure of administration	Cell number/kg or body	Frequency interval	Results	Adverse events
Kidney								
Sun et al. [81]	Prevention of delayed graft function and acute rejection in renal transplantation	21	41.0 ± 11.5	IV	2 × 10^6 cells/kg (before transplantation), 5 × 10^6 (during surgery)	←	No significant improvement	No
Deng et al. [82]	Lupus nephritis	12 MSC, 6 placebo	29 ± 10	IV	1 × 10^8 cells	2 times 1 wk. interval	Not statistically significant improvement	1: leucopenia, pneumonia, subcutaneous abscess, 1: severe pneumonia
Autoimmune diseases								
Wang et al. [83]	Active and refractory SLE	40	17–54	IV	1 × 10^6 cells/kg on day 0 and 7	2	MCR (13 of 40, 32.5%), PCR (11 of 40, 27.5%) during 12 months, although several patients relapse after 6 months	No
Wang et al. [84]	RA	136	46.1	IV	4 × 10^7 cells, 2nd in 3 months later	1 (n = 112) 2 (n = 24)	Decreases of serum TNF-α, IL-6, increase of regulatory T cells. Significant remission for 3–6 months	Mild fever (<38.5°C) without treatment at injection, 6 patients

Authors	Disease	Patients number	Age (range) year	Route and procedure of administration	Cell number/kg or body	Frequency interval	Results	Adverse events
Riordan et al. [85]	Multiple sclerosis	20	41.2 (24–55)	IV	2×10^7 UC-MSC	7 (1–4 days)	Significant improvements of various symptoms. Inactive lesions by MRI in 15/18 patients. (83.3%) after 1 year	Headache, fatigue
Others								
He et al. [86]	Severe sepsis	15 (3 cohorts)	56 (25–72)	IV	1×10^6 cells/kg	1	System clinical outcomes are not changed	No
					2×10^6 cells/kg			
					3×10^6 cells/kg			
Cao et al. [87]	Recurrent intrauterine adhesions	27	35.1 ± 3.8 (27–42)	Loaded onto a collagen scaffold	1×10^7	1	Pregnant, 10 of 26 patients	No

aGVHD, acute graft-versus-host disease; cGVHD, chronic GVHD; HSCT, hematopoietic stem cell transplantation; AA, aplastic anemia; BM, bone marrow; IT, intrathecal injection; AE, adverse event; AML, acute myeloid leukemia; CB, cord blood; UC, umbilical cord; BM, bone marrow; PE, plasma exchange; RA, rheumatoid arthritis; MSCs, mesenchymal stromal cells; DM, diabetes mellitus; FBS, fast blood sugar; ES, ejection fraction; IV, intravenous injection; SC, subcutaneous injection; DM, diabetes mellitus.

Table 3.
Clinical trials using allogeneic umbilical cord-derived mesenchymal stromal cells.

7. Private CB and UC banking for autologous and family use

Recently, there are an increasing number of private CB banks, which have initiated to serve the cryopreservation of UC, i.e., private CB and UC bank. Using private autologous CB, clinical trials for cerebral palsy caused by hypoxic ischemic encephalopathy (HIE) reported their efficacy [88], although the collection of CB is difficult for the baby in such a severe situation of delivery, resulting in the limited application entry. Recently, we obtained the proof of concept that the UC-MSCs attenuated the neurogenic and functional damage caused by intraventricular hemorrhage (IVH) in newborn model mice. Duke University implemented the clinical trial using allogeneic UC tissue-derived cells for the patients with HIE. Allogeneic off-the-shelf UC-MSCs are a promising source; however, we do not know the adverse events such as HLA antibody induction caused by long-term repeated injections of allogeneic cells. Therefore, autologous use of CB and UC is still challenged to be discussed continuously.

8. Conclusion

Although several problems still remain to be dissolved, operation of adequate CB and UC bank should be considered as the provider of cell source for regenerative and immune cell therapy, because of their prominent characteristics and convenient and noninvasive collection.

Acknowledgements

This study was supported by Grants-in-Aids for Scientific Research from the Japan Agency for Medical Research and Development (AMED) (19be0504001h0002). We thank Kamisato A. PhD. for ethical support. We specially thank Takahashi A. MT, Ms. Hori A., Yamamoto Y. MT., Mr. Miharu Y. MT, Ms. Izawa, and MS. Nagaya N., for processing and quality management. We also thank the staff of NTT Medical Center Hospital and Yamaguchi Hospital, Tokyo, for their assistance with the collection of UC and CB.

Author details

Tokiko Nagamura-Inoue[1]* and Fumitaka Nagamura[2]

1 Department of Cell Processing and Transfusion, IMSUT CORD, Research Hospital, The Institute of Medical Science, The University of Tokyo, Tokyo, Japan

2 Division of Advanced Medicine Promotion, The Advanced Clinical Center, The Institute of Medical Science, The University of Tokyo, Tokyo, Japan

*Address all correspondence to: tokikoni@ims.u-tokyo.ac.jp

References

[1] Liu E, Tong Y, Dotti G, et al. Cord blood NK cells engineered to express IL-15 and a CD19-targeted CAR show long-term persistence and potent antitumor activity. Leukemia. 2018;**32**(2):520-531

[2] Gnecchi M, Melo LG. Bone marrow-derived mesenchymal stem cells: Isolation, expansion, characterization, viral transduction, and production of conditioned medium. Methods in Molecular Biology. 2009;**482**:281-294

[3] Bieback K, Kluter H. Mesenchymal stromal cells from umbilical cord blood. Current Stem Cell Research & Therapy. 2007;**2**(4):310-323

[4] Gruber HE, Deepe R, Hoelscher GL, et al. Human adipose-derived mesenchymal stem cells: Direction to a phenotype sharing similarities with the disc, gene expression profiling, and coculture with human annulus cells. Tissue Engineering. Part A. 2010;**16**(9):2843-2860

[5] In 't Anker PS, Scherjon SA, Kleijburg-van der Keur C, et al. Isolation of mesenchymal stem cells of fetal or maternal origin from human placenta. Stem Cells. 2004;**22**(7):1338-1345

[6] Romanov YA, Svintsitskaya VA, Smirnov VN. Searching for alternative sources of postnatal human mesenchymal stem cells: Candidate MSC-like cells from umbilical cord. Stem Cells. 2003;**21**(1):105-110

[7] Stenderup K, Justesen J, Clausen C, Kassem M. Aging is associated with decreased maximal life span and accelerated senescence of bone marrow stromal cells. Bone. 2003;**33**(6):919-926

[8] Rubinstein P, Dobrila L, Rosenfield RE, et al. Processing and cryopreservation of placental/umbilical cord blood for unrelated bone marrow reconstitution. Proceedings of the National Academy of Sciences of the United States of America. 1995;**92**(22):10119-10122

[9] Lapierre V, Pellegrini N, Bardey I, et al. Cord blood volume reduction using an automated system (Sepax) vs. a semi-automated system (Optipress II) and a manual method (hydroxyethyl starch sedimentation) for routine cord blood banking: A comparative study. Cytotherapy. 2007;**9**(2):165-169

[10] Solves P, Mirabet V, Blanquer A, et al. A new automatic device for routine cord blood banking: Critical analysis of different volume reduction methodologies. Cytotherapy. 2009;**11**(8):1101-1107

[11] Park YB, Ha CW, Lee CH, Yoon YC, Park YG. Cartilage regeneration in osteoarthritic patients by a composite of allogeneic umbilical cord blood-derived mesenchymal stem cells and hyaluronate hydrogel: Results from a clinical trial for safety and proof-of-concept with 7 years of extended follow-up. Stem Cells Translational Medicine. 2017;**6**(2):613-621

[12] Nagamura-Inoue T, He H. Umbilical cord-derived mesenchymal stem cells: Their advantages and potential clinical utility. World Journal of Stem Cells. 2014;**6**(2):195-202

[13] Ishige I, Nagamura-Inoue T, Honda MJ, et al. Comparison of mesenchymal stem cells derived from arterial, venous, and Wharton's jelly explants of human umbilical cord. International Journal of Hematology. 2009;**90**(2):261-269

[14] Mennan C, Brown S, McCarthy H, et al. Mesenchymal stromal cells derived from whole human umbilical cord exhibit similar properties to those derived from Wharton's jelly and bone marrow. FEBS Open Bio. 2016;**6**(11):1054-1066

[15] Bharti D, Shivakumar SB, Park JK, et al. Comparative analysis of human Wharton's jelly mesenchymal stem cells derived from different parts of the same umbilical cord. Cell and Tissue Research. 2018;**372**(1):51-65

[16] Shimazu T, Mori Y, Takahashi A, Tsunoda H, Tojo A, Nagamura-Inoue T. Serum- and xeno-free cryopreservation of human umbilical cord tissue as mesenchymal stromal cell source. Cytotherapy. 2015;**17**(5):593-600

[17] Mori Y, Ohshimo J, Shimazu T, et al. Improved explant method to isolate umbilical cord-derived mesenchymal stem cells and their immunosuppressive properties. Tissue Engineering. Part C, Methods. 2015;**21**(4):367-372

[18] Lauterboeck L, Saha D, Chatterjee A, Hofmann N, Glasmacher B. Xeno-free cryopreservation of bone marrow-derived multipotent stromal cells from *Callithrix jacchus*. Biopreservation and Biobanking. 2016;**14**(6):530-538

[19] Ennis J, Gotherstrom C, Le Blanc K, Davies JE. In vitro immunologic properties of human umbilical cord perivascular cells. Cytotherapy. 2008;**10**(2):174-181

[20] Roy S, Arora S, Kumari P, Ta M. A simple and serum-free protocol for cryopreservation of human umbilical cord as source of Wharton's jelly mesenchymal stem cells. Cryobiology. 2014;**68**(3):467-472

[21] He H, Nagamura-Inoue T, Tsunoda H, et al. Stage-specific embryonic antigen 4 in Wharton's jelly-derived mesenchymal stem cells is not a marker for proliferation and multipotency. Tissue Engineering. Part A. 2014;**20**(7-8):1314-1324

[22] Majore I, Moretti P, Stahl F, Hass R, Kasper C. Growth and differentiation properties of mesenchymal stromal cell populations derived from whole human umbilical cord. Stem Cell Reviews. 2011;**7**(1):17-31

[23] Marmotti A, Mattia S, Bruzzone M, et al. Minced umbilical cord fragments as a source of cells for orthopaedic tissue engineering: An in vitro study. Stem Cells International. 2012;**2012**:326813

[24] Kandoi S, Kumar LP, Patra B, et al. Evaluation of platelet lysate as a substitute for FBS in explant and enzymatic isolation methods of human umbilical cord MSCs. Scientific Reports. 2018;**8**(1):12439

[25] Kikuchi-Taura A, Taguchi A, Kanda T, et al. Human umbilical cord provides a significant source of unexpanded mesenchymal stromal cells. Cytotherapy. 2012;**14**(4):441-450

[26] Salehinejad P, Alitheen NB, Ali AM, et al. Comparison of different methods for the isolation of mesenchymal stem cells from human umbilical cord Wharton's jelly. In Vitro Cellular & Developmental Biology. Animal. 2012;**48**(2):75-83

[27] Das R, Roosloot R, van Pel M, et al. Preparing for cell culture scale-out: Establishing parity of bioreactor- and flask-expanded mesenchymal stromal cell cultures. Journal of Translational Medicine. 2019;**17**(1):241

[28] Dominici M, Le Blanc K, Mueller I, et al. Minimal criteria for defining multipotent mesenchymal stromal cells. The International Society for Cellular Therapy position statement. Cytotherapy. 2006;**8**(4):315-317

[29] Horwitz EM, Le Blanc K, Dominici M, et al. Clarification of the nomenclature for MSC: The International Society for Cellular Therapy position statement. Cytotherapy. 2005;**7**(5):393-395

[30] Wang JF, Wang LJ, Wu YF, et al. Mesenchymal stem/progenitor cells in human umbilical cord blood as support for ex vivo expansion of CD34(+) hematopoietic stem cells and for chondrogenic differentiation. Haematologica. 2004;**89**(7):837-844

[31] Karahuseyinoglu S, Cinar O, Kilic E, et al. Biology of stem cells in human umbilical cord stroma: In situ and in vitro surveys. Stem Cells. 2007;**25**(2):319-331

[32] Huang P, Lin LM, Wu XY, et al. Differentiation of human umbilical cord Wharton's jelly-derived mesenchymal stem cells into germ-like cells in vitro. Journal of Cellular Biochemistry. 2010;**109**(4):747-754

[33] Weiss ML, Anderson C, Medicetty S, et al. Immune properties of human umbilical cord Wharton's jelly-derived cells. Stem Cells. 2008;**26**(11):2865-2874

[34] Chiossone L, Conte R, Spaggiari GM, et al. Mesenchymal stromal cells induce peculiar alternatively activated macrophages capable of dampening both innate and adaptive immune responses. Stem Cells. 2016;**34**(7):1909-1921

[35] Deng Y, Zhang Y, Ye L, et al. Umbilical cord-derived mesenchymal stem cells instruct monocytes towards an IL10-producing phenotype by secreting IL6 and HGF. Scientific Reports. 2016;**6**:37566

[36] Cutler AJ, Limbani V, Girdlestone J, Navarrete CV. Umbilical cord-derived mesenchymal stromal cells modulate monocyte function to suppress T cell proliferation. Journal of Immunology. 2010;**185**(11):6617-6623

[37] Selleri S, Bifsha P, Civini S, et al. Human mesenchymal stromal cell-secreted lactate induces M2-macrophage differentiation by metabolic reprogramming. Oncotarget. 2016;7(21):30193-30210

[38] Seshareddy K, Troyer D, Weiss ML. Method to isolate mesenchymal-like cells from Wharton's jelly of umbilical cord. Methods in Cell Biology. 2008;**86**:101-119

[39] Zhou C, Yang B, Tian Y, et al. Immunomodulatory effect of human umbilical cord Wharton's jelly-derived mesenchymal stem cells on lymphocytes. Cellular Immunology. 2011;**272**(1):33-38

[40] (FACT) FFTAOCT. International Standards For Cord Blood Collection, Banking, And Release For Administration. 7th ed2020

[41] Brunstein CG, Miller JS, McKenna DH, et al. Umbilical cord blood-derived T regulatory cells to prevent GVHD: Kinetics, toxicity profile, and clinical effect. Blood. 2016;**127**(8):1044-1051

[42] Kellner JN, Delemarre EM, Yvon E, et al. Third party, umbilical cord blood derived regulatory T-cells for prevention of graft versus host disease in allogeneic hematopoietic stem cell transplantation: Feasibility, safety and immune reconstitution. Oncotarget. 2018;**9**(86):35611-35622

[43] Zhu H, Poon W, Liu Y, et al. Phase I-II clinical trial assessing safety and efficacy of umbilical cord blood mononuclear cell transplant therapy of chronic complete spinal cord injury. Cell Transplantation. 2016;**25**(11):1925-1943

[44] Shah N, Li L, McCarty J, et al. Phase I study of cord blood-derived natural killer cells combined with autologous stem cell transplantation in multiple myeloma. British Journal of Haematology. 2017;**177**(3):457-466

[45] Lv YT, Zhang Y, Liu M, et al. Transplantation of human cord blood mononuclear cells and umbilical cord-derived mesenchymal stem cells in autism. Journal of Translational Medicine. 2013;**11**:196

[46] Dolstra H, Roeven MWH, Spanholtz J, et al. Successful transfer of umbilical cord blood CD34(+) hematopoietic stem and progenitor-derived NK cells in older acute myeloid Leukemia patients. Clinical Cancer Research. 2017;**23**(15):4107-4118

[47] Park EH, Lim HS, Lee S, et al. Intravenous infusion of umbilical cord blood-derived mesenchymal stem cells in rheumatoid arthritis: A phase Ia clinical trial. Stem Cells Translational Medicine. 2018;7(9):636-642

[48] Laskowitz DT, Bennett ER, Durham RJ, et al. Allogeneic umbilical cord blood infusion for adults with ischemic stroke: Clinical outcomes from a phase I safety study. Stem Cells Translational Medicine. 2018;7(7):521-529

[49] Huang L, Zhang C, Gu J, et al. A randomized, placebo-controlled trial of human umbilical cord blood mesenchymal stem cell infusion for children with cerebral palsy. Cell Transplantation. 2018;**27**(2):325-334

[50] Kim HS, Lee JH, Roh KH, Jun HJ, Kang KS, Kim TY. Clinical trial of human umbilical cord blood-derived stem cells for the treatment of moderate-to-severe atopic dermatitis: Phase I/IIa studies. Stem Cells. 2017;**35**(1):248-255

[51] Wu KH, Chan CK, Tsai C, et al. Effective treatment of severe steroid-resistant acute graft-versus-host disease with umbilical cord-derived mesenchymal stem cells. Transplantation. 2011;**91**(12):1412-1416

[52] Si Y, Yang K, Qin M, et al. Efficacy and safety of human umbilical cord derived mesenchymal stem cell therapy in children with severe aplastic anemia following allogeneic hematopoietic stem cell transplantation: A retrospective case series of 37 patients. Pediatric Hematology and Oncology. 2014;**31**(1):39-49

[53] Wu Y, Wang Z, Cao Y, et al. Cotransplantation of haploidentical hematopoietic and umbilical cord mesenchymal stem cells with a myeloablative regimen for refractory/relapsed hematologic malignancy. Annals of Hematology. 2013;**92**(12):1675-1684

[54] Wu Y, Cao Y, Li X, et al. Cotransplantation of haploidentical hematopoietic and umbilical cord mesenchymal stem cells for severe aplastic anemia: Successful engraftment and mild GVHD. Stem Cell Research. 2014;**12**(1):132-138

[55] Fu Y, Wang Q, Zhou J, et al. Reduced intensity conditioning and co-transplantation of unrelated peripheral stem cells combined with umbilical cord mesenchymal stem/stroma cells for young patients with refractory severe aplastic anemia. International Journal of Hematology. 2013;**98**(6):658-663

[56] Gao L, Zhang Y, Hu B, et al. Phase II Multicenter, randomized, double-blind controlled study of efficacy and safety of umbilical cord-derived mesenchymal stromal cells in the prophylaxis of chronic graft-versus-host disease after HLA-haploidentical stem-cell transplantation. Journal of Clinical Oncology. 2016;**34**(24):2843-2850

[57] Zhu L, Wang Z, Zheng X, et al. Haploidentical hematopoietic stem cell transplant with umbilical cord-derived multipotent mesenchymal cell infusion for the treatment of high-risk acute leukemia in children. Leukemia & Lymphoma. 2015;**56**(5):1346-1352

[58] Pan Y, Wang X, Wang C, et al. Extensive bone marrow necrosis resolved by allogeneic umbilical cord blood mesenchymal stem cell transplantation in a chronic myeloid leukemia patient. Bone Marrow Transplantation. 2015;**50**(9):1265-1268

[59] Wang S, Cheng H, Dai G, et al. Umbilical cord mesenchymal stem cell transplantation significantly improves neurological function in patients with sequelae of traumatic brain injury. Brain Research. 2013;**1532**:76-84

[60] Jin JL, Liu Z, Lu ZJ, et al. Safety and efficacy of umbilical cord mesenchymal stem cell therapy in

hereditary spinocerebellar ataxia. Current Neurovascular Research. 2013;**10**(1):11-20

[61] Wang X, Hu H, Hua R, et al. Effect of umbilical cord mesenchymal stromal cells on motor functions of identical twins with cerebral palsy: Pilot study on the correlation of efficacy and hereditary factors. Cytotherapy. 2015;**17**(2):224-231

[62] Guan LX, Guan H, Li HB, et al. Therapeutic efficacy of umbilical cord-derived mesenchymal stem cells in patients with type 2 diabetes. Experimental and Therapeutic Medicine. 2015;**9**(5):1623-1630

[63] Hu J, Yu X, Wang Z, et al. Long term effects of the implantation of Wharton's jelly-derived mesenchymal stem cells from the umbilical cord for newly-onset type 1 diabetes mellitus. Endocrine Journal. 2013;**60**(3):347-357

[64] Cai J, Wu Z, Xu X, et al. Umbilical cord mesenchymal stromal cell with autologous bone marrow cell transplantation in established type 1 diabetes: A pilot randomized controlled open-label clinical study to assess safety and impact on insulin secretion. Diabetes Care. 2016;**39**(1):149-157

[65] Kong D, Zhuang X, Wang D, et al. Umbilical cord mesenchymal stem cell transfusion ameliorated hyperglycemia in patients with type 2 diabetes mellitus. Clinical Laboratory. 2014;**60**(12):1969-1976

[66] Cai J, Wu Z, Huang L, et al. Cotransplantation of bone marrow mononuclear cells and umbilical cord mesenchymal stem cells in avascular necrosis of the femoral head. Transplantation Proceedings. 2014;**46**(1):151-155

[67] Can A, Ulus AT, Cinar O, et al. Human umbilical cord mesenchymal stromal cell transplantation in myocardial ischemia (HUC-HEART Trial). A study protocol of a phase 1/2, controlled and randomized trial in combination with coronary artery bypass grafting. Stem Cell Reviews and Reports. 2015;**11**(5):752-60

[68] Zhao XF, Xu Y, Zhu ZY, Gao CY, Shi YN. Clinical observation of umbilical cord mesenchymal stem cell treatment of severe systolic heart failure. Genetics and Molecular Research. 2015;**14**(2):3010-3017

[69] Li X, Hu YD, Guo Y, et al. Safety and efficacy of intracoronary human umbilical cord-derived mesenchymal stem cell treatment for very old patients with coronary chronic total occlusion. Current Pharmaceutical Design. 2015;**21**(11):1426-1432

[70] Musialek P, Mazurek A, Jarocha D, et al. Myocardial regeneration strategy using Wharton's jelly mesenchymal stem cells as an off-the-shelf 'unlimited' therapeutic agent: Results from the acute myocardial infarction first-in-man study. Postępy w Kardiologii Interwencyjnej. 2015;**11**(2):100-107

[71] Bartolucci J, Verdugo FJ, Gonzalez PL, et al. Safety and efficacy of the intravenous infusion of umbilical cord mesenchymal stem cells in patients with heart failure: A phase 1/2 randomized controlled trial (RIMECARD trial [Randomized clinical trial of intravenous infusion umbilical cord mesenchymal stem cells on cardiopathy]). Circulation Research. 2017;**121**(10):1192-1204

[72] Xue HL, Zeng WZ, Wu XL, et al. Clinical therapeutic effects of human umbilical cord-derived mesenchymal stem cells transplantation in the treatment of end-stage liver disease. Transplantation Proceedings. 2015;**47**(2):412-418

[73] Wang L, Li J, Liu H, et al. A pilot study of umbilical cord-derived mesenchymal stem cell transfusion in patients with primary biliary cirrhosis. Journal of Gastroenterology and Hepatology. 2013;**28**(Suppl 1):85-92

[74] Shi M, Liu Z, Wang Y, et al. A pilot study of mesenchymal stem cell therapy for acute liver allograft rejection. Stem Cells Translational Medicine. 2017;**6**(12):2053-2061

[75] Liang J, Zhang H, Zhao C, et al. Effects of allogeneic mesenchymal stem cell transplantation in the treatment of liver cirrhosis caused by autoimmune diseases. International Journal of Rheumatic Diseases. 2017;**20**(9):1219-1226

[76] Zhang YC, Liu W, Fu BS, et al. Therapeutic potentials of umbilical cord-derived mesenchymal stromal cells for ischemic-type biliary lesions following liver transplantation. Cytotherapy. 2017;**19**(2):194-199

[77] Xu WX, He HL, Pan SW, et al. Combination treatments of plasma exchange and umbilical cord-derived mesenchymal stem cell transplantation for patients with Hepatitis B virus-related acute-on-chronic liver failure: A clinical trial in China. Stem Cells International. 2019;**2019**:4130757

[78] Zhang J, Lv S, Liu X, Song B, Shi L. Umbilical cord mesenchymal stem cell treatment for Crohn's disease: A randomized controlled clinical trial. Gut Liver. 2018;**12**(1):73-78

[79] Hu J, Zhao G, Zhang L, et al. Safety and therapeutic effect of mesenchymal stem cell infusion on moderate to severe ulcerative colitis. Experimental and Therapeutic Medicine. 2016;**12**(5):2983-2989

[80] Hashemi SS, Mohammadi AA, Kabiri H, et al. The healing effect of Wharton's jelly stem cells seeded on biological scaffold in chronic skin ulcers: A randomized clinical trial. Journal of Cosmetic Dermatology. 2019;**18**(6):1961-1967

[81] Sun Q, Huang Z, Han F, et al. Allogeneic mesenchymal stem cells as induction therapy are safe and feasible in renal allografts: Pilot results of a multicenter randomized controlled trial. Journal of Translational Medicine. 2018;**16**(1):52

[82] Sun L, Wang D, Liang J, et al. Umbilical cord mesenchymal stem cell transplantation in severe and refractory systemic lupus erythematosus. Arthritis and Rheumatism. 2010;**62**(8):2467-2475

[83] Wang D, Zhang H, Liang J, et al. Allogeneic mesenchymal stem cell transplantation in severe and refractory systemic lupus erythematosus: 4 years of experience. Cell Transplantation. 2013;**22**(12):2267-2277

[84] Wang L, Wang L, Cong X, et al. Human umbilical cord mesenchymal stem cell therapy for patients with active rheumatoid arthritis: Safety and efficacy. Stem Cells and Development. 2013;**22**(24):3192-3202

[85] Riordan NH, Morales I, Fernandez G, et al. Clinical feasibility of umbilical cord tissue-derived mesenchymal stem cells in the treatment of multiple sclerosis. Journal of Translational Medicine. 2018;**16**(1):57

[86] He X, Ai S, Guo W, et al. Umbilical cord-derived mesenchymal stem (stromal) cells for treatment of severe sepsis: A phase 1 clinical trial. Translational Research. 2018;**199**:52-61

[87] Cao Y, Sun H, Zhu H, et al. Allogeneic cell therapy using umbilical cord MSCs on collagen scaffolds for patients with recurrent uterine adhesion: A phase I clinical trial. Stem Cell Research & Therapy. 2018;**9**(1):192

[88] Cotten CM, Murtha AP, Goldberg RN, et al. Feasibility of autologous cord blood cells for infants with hypoxic-ischemic encephalopathy. The Journal of Pediatrics. 2014;**164**(5):973-979. e971

Section 2

Ethical and Legal Aspects

Chapter 4

Patentability of the Human Embryonic Stem Cell Lines: A Legal and Ethical Aspect

Tansu Sayar Kanyış, Ezgi Arslan and Oğuzhan Kanyış

Abstract

In this study, patentability of the human embryonic stem cell lines has discussed in the legal and ethical perspectives. In vitro human embryonic stem cells can be defined as body parts that are departed from the body. Human embryonic stem cell lines are constituted of differentiated self-renewal pluripotent stem cells, which means they have no characteristics to become a human-being. However, interpreting the terms like human embryo and right to property widely can cause the human embryonic stem cell lines are misunderstood as unpatentable. For our point of view, giving the human embryo the protections of both personal rights of the donor and the right to property of the owner of the invention does not reduce the legal/moral status of the human embryo. Besides, the obligations which these rights imposes to their owners, such as the principle of human dignity and prohibition of financial gain can protect the human embryo in a better way.

Keywords: human embryonic stem cell lines, patentability, human dignity, personal rights, right to property, prohibition of financial gain, law, ethics

1. Introduction

Human embryonic stem cell researches (HESCRs) has been widely discussed in different perspectives. In this chapter, the issue will be considered from the perspective of the patentability of the human embryonic stem cell lines (HESCLs). Embryonic stem cells are derived from the inner cell mass of an embryo in the blastocyst stage. Embryonic stem cells are "pluripotent" cells, which means, they are, technically, differentiable into a wide range of cell and tissue types [1]. The researches on human embryonic stem cells have the potential to discover and develop treatments for a variety of diseases including Alzheimer's disease, diabetes, neurodegenerative disorders, heart diseases, Parkinson's disease, or anemia [1, 2]. HESCLs make scientific researches and stem cell treatments possible by producing differentiated self-renewal pluripotent stem cells. However, conflicts of interest arise between patients seeking treatment and human embryos from the fact that, while harvesting HESCs, the human embryo is destroyed.

The subject of patentability of the HESCLs has also been discussed because of the legal prohibition of financial gain on the human body and its parts and the principle of human dignity [3]. It is not possible to accept human embryo and HESCLs as a property that is suitable for industrial applicability when the concepts are regarded with their traditional definitions. The principle of human dignity and the prohibition of financial gain on the human body and its parts prevent such acceptance. However, the human embryo and the stem cells derived from it have a *sui generis* legal and moral status. Besides, in the patentability of the HESCLs issue, the patentable thing is not the human embryo itself, but the pluripotent cell lines differentiated into self-renewal pluripotent stem cells and the method of the process.

In the first part of this chapter, the *sui generis* legal status of the human embryo will be discussed by interpreting the legal status of the human body parts. In the second part, conditions of patentability will be analyzed from the perspective of the patentability of the HESCLs. Finally, in the third part of this chapter, the patentability of the HESCLs will be viewed in the moral and ethical aspects.

2. Legal status of the human embryo

It is possible to make an interpretation based on the legal status of the human body parts to understand the legal status of the in vitro human embryonic stem cells. Such an interpretation needs a few definitions and premises:

1. We can define human body parts as the body parts of a person.

2. Human body parts can be divided into two such as inter-body parts (which are in a unity with the body) and the body parts that are departed from the body [4].

3. It is possible to divide human cells into two such as somatic cells and germ cells.

4. A human embryo is a being that is composed of germ cells of women (OVAs) and men (sperms).

5. A human embryo can be in vivo (in the body of the pregnant woman) or in vitro (in the outside of a woman's body, probably in a Petri tube).

6. Human embryonic stem cells are derived from the inner cell mass of an embryo in the blastocysts stage [1].

7. Embryonic stem cells are "pluripotent" cells, which means, they are, technically, differentiable into a wide range of cell and tissue types [1]. However, they are not "totipotent" cells. A basic meaning of this fact is: they cannot differentiate into a human embryo; they can only differentiate into specific types of cells and tissues [5].

8. Since we agree with the general acceptance of the human embryo has no legal personality, we are in the view of human embryonic stem cells of an in vivo human embryo is within the body parts of the pregnant woman.

However, things are much more complicated in the subject of the in vitro embryo. Does it belong to the donor of the ova or the donor of the sperm or both of them? Does it belong to the patient of assisted reproduction treatment? What happens if it is donated for the researches? Does it belong to the researcher? What does "to belong to" mean for human body parts? Does it mean ownership? Is it possible to accept human body parts as properties?

According to the traditional definition, the right to property is a right, which gives its owner the authority to use, to enjoy the fruits and the ownership. If one of these three authorities not legally enjoyable for a person, then that person has no right to property on that property or that thing is not subject to the right to property. Authority to ownership means legal availability to buy and sell. Authority to ownership is not enjoyable for body parts because the prohibition of financial gain on the human body and its parts is a fundamental principle, which is accepted worldwide in both international and domestic regulations. Article 21 of the Convention on Human Rights and Biomedicine (CETS No. 164) with the title of "*prohibition of financial gain*" is a good example of such regulations by saying "*The human body and its parts shall not, as such, give rise to financial gain*" [6]. Another example is Article 3 of the European Union Charter of Fundamental Rights. According to the article with the title of "Right to the integrity of the person,"

1. *"Everyone has the right to respect for his or her physical and mental integrity.*

2. *In the fields of medicine and biology, the following must be respected in particular:*

- *the free and informed consent of the person concerned, according to the procedures laid down by law,*

- *the prohibition of eugenic practices, in particular those aiming at the selection of persons,*

- *the prohibition on making the human body and its parts as such a source of financial gain,*

- *the prohibition of the reproductive cloning of human beings"* [7].

Because of the prohibition of financial gain on the human body and its parts, according to the general view in the French, German and Turkish doctrine, germ cells that are departed from the body cannot be considered as property, and giving harm to these cells causes a violation of personal rights, not of the right to property [4, 8–13]. While some of the authors claim that protecting human embryos is possible only with the personal rights of the donors [8–10] others emphasize its *sui generis* legal status [11–13].

As it is seen while the doctrine agrees with the view of human embryos have a sui generis legal status and can be protected with the donor's personal rights, there is no unity in the view of it is also possible to protect the right to property to the human embryos. As we mentioned before, it is possible to make an interpretation based on the legal status of the human body parts to understand the legal status of the in vitro human embryonic stem cells. Because human germ cells are about reproductive rights, more specific protection is provided for the germ cells than that of the somatic cells. Because the human embryo has characteristics of both the donors' and the embryo itself and also has the potential to develop into a living human, more specific protection is provided for in vivo human embryo (for example with abortion laws) than that of the germ cells. On the other hand, such a more specific protection is not provided to in vitro human embryo, although it has the same characteristics as in vivo human embryo. From our point of view, protecting it with the right to property besides the protection of personal rights can give the more specific protection the in vitro human embryo needs.

3. The patentability of HESCLs from the legal aspect

A patent is a legal document, which provides a right to enjoy the innovation's owner and prevent third parties from violating the rights of the owner. The owner of the patent has the opportunity to declare the rights related to the patent to third parties [14]. By giving patents, states aim to encourage scientists to make science. An innovation that is given patent improves the scientific development of societies by presenting the innovation to the memory of the society. States target community development by authorizing exclusive competence to the owner of the innovation [15].

Legislations which affect the patentability of HESCLs usually regulate the patentability of the biotechnological inventions field, in which the applications about the HESCLs take part. Unfortunately there is no legislation regulates the patentability of the HESCLs specificly. For this reason, rules for the patentability of the HESCLs are reachable by interpreting legislations on the biotechnological inventions. Therefore, in this part of the syudy, patentability criteria and international regulations on the biotechnological inventions will be examined in the first two subtitles. Then, in the third subtitle of this part, patentability of the HESCLs will be specificly considered.

3.1 The relationship between biotechnological inventions and patentability criteria

Patents are qualified with three criteria. To an application is being patented, it has to be novel and inventive and have industrial applicability [16]. These criteria are also needed for the patent application in the field of biotechnology. However, because of the sui generis characteristics of biotechnological researches, some differences appear. In the paragraphs below, the affection of the sui generis characteristics of the biotechnology field on the meaning of the patentability criteria will be explained.

Novelty, which is the first characteristic of patentability, means going beyond the state of art in the field of biotechnology [17]. To a research in biotechnology is beyond the state of art, it has to be about a technic that is more developed than the technic that was known before. It is not enough the research itself has the ethical values that are determined by the international organizations, the novel technic should also be compatible with these values. The qualification on the novelty in the field of Biotechnology is not far more different from the qualification on the novelty in the other fields. However, in the field of biotechnology, there has to be a specific qualification, since in these innovations the materials existed in nature are mentioned. The fact that proteins, genes, enzymes, and such materials are already existed in nature and in the researches such materials are processed, the characteristic of novelty does not appear traditionally. For this reason, in the field of biotechnology studying more on material or finding the different physical characteristics and forms of the materials do not eliminate the innovations characteristic of novelty. Also the fact that its benefits are already known does not eliminate the characteristic of novelty [18]. However, if the subject of the patent application has the same technic with the technics that are already known, it is accepted that the application has not got the characteristic of novelty, without considering its production method.

The second criteria for a patent application is its inventiveness. A patent application should be inventive. In this context, firstly, the target of the application should be determined. If there is a determined target for the application and if this target is reachable with the estimated theory and the existing information for the application, it is possible to say that there is an innovation. Another matter, which proves that there is an invention, is the application is capable of fulfilling the existing needs. Besides, there should be a reasonable expectation for the technic to be "obviously" successful, which means each time the technic has used the consequences should be the same. No application that has obvious consequences can be patented [19].

PharmaStem v Viacell Case of United States District Court of Delaware is an example of the conflicts on inventiveness in Europe. The case is about an invention on an isolated DNA molecule, which codes human tissue plasminogen activator (t-PA). The court has decided that producing human t-PA by using human recombinant DNA technology is an obvious consequence for any expert in the field. The court says that, oligonucleotide probing was a known technic, any expert could reach the consequence of the invention that seeking for the protection of patent because choosing oligonucleotide probs did not need a high level of skill and experience. According to the court, the monopoly rights given to the patent owners provide much more than a prize given for winning the race of recombinant expression of the gene. For this reason, the court says that the invention seeking patent protection could not succeed in the criteria of inventiveness. It is seen that the court has emphasized the obviousness of the isolation methods of DNA, not the structure of the molecule, and reached the conclusion of the decision should be made by considering the creativity of the method, not by considering the speed of the method's application [16, 20]. As is seen, the criteria of inventiveness qualified specially in biotechnological researches.

The last criteria for a patent application is industrial applicability. Determination of the industrial applicability can be hard for a biotechnological invention because, in general, it is not as clear as for an invention on a gene or a protein sequence, as for the inventions in other fields of science. For example, in genetic researches, short DNA sequences' or expressed sequence tags are used as probs. However, some quarters claim that expressed sequence tags do not have enough benefits for patentability [16].

A patent application, when it is considered with the knowledge that is widely known, should include a real industrial applicability expectation, instead of a completely theoretical probability of industrial applicability. Without a clear description related to its method or describing the method without pointing out one of its practical benefits, it is not enough for the industrial applicability criteria to succeed. It is also not enough to relate the structure to some reachable but undetermined theoretical aims for the industrial applicability criteria to succeed. However, having no experiment or laboratory data related to the method of the application seeking patent protection does not show that it is not reaching the industrial applicability criteria. Criteria of industrial applicability cannot be dependent on experiments or laboratory data. It is enough to give a reasonable and reliable benefit or estimated data. Reliability can be supported by the possible information that can appear later. Because laboratory reports, expert opinions, and clinical trials related to the invention increase the reliability of the applicability of the invention, they support the industrial applicability criteria indirectly [21].

It is possible having no industrial applicability criteria for one part of the invention to affect the industrial applicability criteria of its other part. For example, when a part of the invention is about receptor, if the receptor has no industrial applicability, the agonist (compounds forming reaction in the cell by connecting to the cell receptors) related to the receptor also have no industrial applicability. Likewise, a method, which defines an agonist related to a receptor, has no industrial applicability, either. On the other hand, it would not be possible to say that the receptor, agonists, and the method of defining the agonists do not have the criteria of industrial applicability if, for example, it is clarified with some in vivo or in vitro data in the description that it is about the treatment of obesity [16]. To sum up, the situation of a part of the application is the lack of the criteria of industrial applicability does not mean that the whole application is not patentable. Hence, the effect of technological developments on the criteria of industrial applicability increases in each passing day.

3.2 Patentability of biotechnological inventions in international conventions

In this part of the study, international regulations on the patentability will be discussed. In this context, The European Patent Convention and The Agreement on Trade-Related Aspects of Intellectual Property Rights are important for regulating general criteria for patentability and its exceptions. Directive on the Legal Protection of Biotechnological Inventions is especially related to the subject of our study by regulating the general rules and exceptions of the patentability of stem cells.

The European Patent Convention [22] (Convention) makes giving valid patents in the state parties possible. Although there is no specific regulation on the patentability of biotechnological inventions, it is possible to conclude by analyzing its general provisions. To an invention is patentable in the context of the Convention, it should have the criteria of novelty, inventiveness, and applicability in industry. In the article 53 of the Convention, the unpatentable inventions are counted. According to article 53/a the patent protection cannot be provided to the inventions which violate ordre public and morality. In the article 53/b, it is regulated that the patent protection cannot also be provided to "*plant or animal varieties or essentially biological processes for the production of plants or animals*". Finally, according to the article 53/c European patents cannot be granted to "*methods for treatment of the human or animal body by surgery or therapy and diagnostic methods practiced on the human or animal body*".

The Agreement on Trade-Related Aspects of Intellectual Property Rights (TRIP's) is another international agreement on patentability. In the article 27 of TRIP's limits for the patentability of inventions are regulated. The article is important for having wide coverage for the protection granted with patentability. According to the article, nearly all inventions, without considering its place, its ability to export or its technological area, are patentable and thereby they can be released into commercial circulation like the other trade objects. However, there are some exceptions to this wide regulation, too. The most important exception which is also essential for the biotechnological inventions is "ordre public and morality". This regulation of the TRIP's has parallels with article 53 of the Convention. Both regulations need reconsidering because the concepts of ordre public and morality are open to interpretation. In this interpretation biotechnological developments should also be considered [15].

Another international agreement in which the patentability of biotechnological researches and especially stem cells are regulated in Directive 98/44/EC of The European Parliament and of The Council of 6 July 1998 on The Legal Protection of Biotechnological Inventions (Directive). The Directive is criticized for including uncertainties and gaps. The Directive's attribution of being blocked off in a matter of confliction in which it should be clarified causes difficulties in interpretation. For this reason, modern technologies and necessities should contribute to the interpretation of the Directive.

The term "biotechnology" which appears in the title of the Directive and constitutes its target means, application of industrial and commercial processes on a biological material (living cells and microorganisms) with scientific methods.

This definition matches up with the definition in article 2 of the United Nations Convention on Biological Diversity [23] dated 1992. Because the investments in the field of biotechnology have high risks, the investors and owners of the inventions need to have legal protection and the preventions against them to be removed. So, the target is regulating the limitations on biotechnological material, especially on scientific researches related to the human body; not constituting a new and special kind of patent for the field of biotechnology. By regulating the limitations, it is aimed at both supporting biotechnological developments and reaching a level of development proper to public order and morality which is regulated in domestic and international regulations [24].

In the Directive, while the prohibited and conditioned subjects are determined, it is also regulated whether biotechnological inventions are patentable or not. According to article 3(2) of the Directive, biological materials are patentable if they are isolated from their natural environment. The article regulates that, "*The human body, at the various stages of its formation and development, and the simple discovery of one of its elements, including the sequence or partial sequence of a gene, cannot constitute patentable inventions*". However in the second paragraph of the same article says that "*an element isolated from the human body or otherwise produced by means of a technical process, including the sequence or partial sequence of a gene, may constitute a patentable invention, even if the structure of that element is identical to that of a natural element*". According to article 6 of the Directive,

> "1. *Inventions shall be considered unpatentable where their commercial exploitation would be contrary to ordre public or morality; however, exploitation shall not be deemed to be so contrary merely because it is prohibited by law or regulation.*
>
> 2. *On the basis of paragraph 1, the following, in particular, shall be considered unpatentable:*
>
> a. *processes for cloning human beings;*
>
> b. *processes for modifying the germ line genetic identity of human beings;*
>
> c. *uses of human embryos for industrial or commercial purposes;*
>
> d. *processes for modifying the genetic identity of animals which are likely to cause them suffering without any substantial medical benefit to man or animal, and also animals resulting from such processes.*"

As it is seen, ordre public and morality are also regulated in the Directive as limits of patentability of biotechnological inventions. Cloning human beings, modifying the germ line genetic identity of human beings, uses of human embryos for industrial or commercial purposes, and modifying the genetic identity of animals are accepted to be out of the patent protection.

The problem in the patentability of HESCRs is the Directive's tendency to use the term "human embryo". However, the term "human embryo" is not defined in the Directive. This situation causes the term human embryo to be widely interpreted as if it is the same thing with pluripotent cells of the human embryo. In fact, embryonic stem cells are "pluripotent" cells, which means, they are, technically, differentiable into a wide range of cell and tissue types [1]. However, they are not "totipotent" cells. A basic meaning of this fact is: They cannot differentiate into a human embryo; they can only differentiate into specific types of cells and tissues [5]. So, in the HESCRs pluripotent stem cells are used, processed, and human embryonic stem cell lines are produced. Neither pluripotent stem cells nor the human embryonic stem cell lines are the same thing with human embryos. Besides, it is not possible for both of them to be used for producing or cloning human embryos. Pretending as if they are the same thing with the human embryos causes the patentability of human embryonic stem cell lines to be trapped into the discussion of the destruction of human embryos for research purposes which is another, but not the same, side of the story. As it is seen, when the knowledge learned with modern technology is not gathered with the regulations written before by the way of interpretation, the regulations cannot succeed in their aims.

As an example of the problem mentioned above, the European Court of Justice has decided in its Oliver Brüstle v Greenpeace eV. case (dated 18.10.2011, numbered c-34/10) that an application is unpatentable by interpreting the term of human embryo widely, in the context of article 6(2)c of the Directive. When the case is analyzed, it is seen that the Court has considered the principle of human dignity and the prohibition of commercial use in its decision. However, interpreting every patent application related to HESCRs within the contexts of human dignity and commercial use would cause a broad interpretation which is contrary to the aim of the Directive. Surely, human dignity is a fundamental principle, which should be considered in every legislating and interpreting process and the Directive serves for human dignity with all of its existence. However, both the Directive and the other international regulations on the matter also aim to support the researches to find treatments for severe and life quality reductive diseases such as Alzheimer's, pancreas cancer, blindness, and Parkinson's. For this reason, terms like human dignity, human embryo, or commercial use should not be interpreted in a way to cause go far from the total target. Preventing inventions with cruel commercial and industrial purposes and presentation of such inventions to the use is a noble cause which should be granted with legal regulations. However, steps to this cause should not be stopped from preventing human suffering. HSCRs need large amounts of economic resources. In order to that economical resources are granted, and researches are developed, patents are very important. The fact that an invention has an economical value should not shadow its necessity for the sake of humankind. However, to maintain the necessities of human dignity, the opportunity for each human being to reach such inventions in an equal way should be granted in the legal regulations.

3.3 The effect of the protection of the absolute rights on the patentability of the HESCLs

The sui generis legal status of human embryo basis on the characteristics of the ability to develop into a living human being, a person [4]. Because of this sui generis legal status and developments in science and medicine especially in the field of stem cell science, new interpretations on the right to property appeared. According to such interpretations, it is possible to provide human embryos the protection of both personal rights and the right to property. Providing the protection of the right to property besides the protection of personal rights do not decrease the legal status of the human embryo that of the property [4, 25]. On the contrary, providing extra protection increases the legal status of the human embryo [26, 27]. However, if these cells will be used as a continuation tool of the donor's personality with the assisted reproductive technologies' help it is better not to entitle them the legal status of the property [25]. In order to the subject of patentability of HESCRs is understand better, in this part of the study we will clarify the terms of personal rights and right to property, which are the branches of the absolute rights, the subjects of these rights, and the necessities of these rights for the patentability.

Absolute rights are the rights that give its owner the broadest power over the owner's own personality and the property they are subject to. Everyone must respect these rights and obey the limits drawn by them. In this respect, absolute rights are powerful rights that can be claimed against every third person. Personal rights are a type of absolute rights, which arise from being a person and they are a tool for the protection of the human embryo. With these rights, which cannot be waived, transferred be converted into money and can only be subject to a claim for compensation in case of violation, it is desired to protect the person's material and moral existence.

As it is mentioned before, HESCs are body parts that can be departed from the body and they include their donors' personal data. For this reason, personal rights of the donors' need to be protected in the HESCRs. It is important to protect the personal data with personal rights against the interventions of the third parties. In his scope, personal data of the donors are under the protection against revealment and unauthorized use with the domestic and international regulations. Oviedo Convention [28] is a good example for such international regulations. According to article 5 of the Oviedo Convention;

> *"An intervention in the health field may only be carried out after the person concerned has given free and informed consent to it.*
>
> *This person shall beforehand be given appropriate information as to the purpose and nature of the intervention as well as on its consequences and risks.*
>
> *The person concerned may freely withdraw consent at any time."*

Article 22 of the Convention is also important for the subject of this study. According to the article 22, "*When in the course of an intervention any part of a human body is removed, it may be stored and used for a purpose other than that for which it was removed, only if this is done in conformity with appropriate information and consent procedures.*" If we interpret the articles for the case of HESCRs, we can say that before the human embryo is or the ova and sperms to constitute the human embryo are taken from the donors, the informed consent of the donors should be taken. The informed consent should include the purpose, process, and possible conclusions of the intervention. If the purpose of the intervention was, for example, treatment of assisted reproduction but the purpose has changed after having surplus embryos and the experts are willing to use the surplus embryos for HESCRs, they should take a new informed consent from the donors and clear the target, process and possible conclusions of their new purpose, too. As it is seen, it is not possible to accept the patent application for HESCRs if the application includes HESCs that is taken without the donor's informed consent. And if such research is done, the personal data of the donor will be protected with his/her personal rights [29].

As the personal rights of a person are protected against the interventions of the third parties, they are also protected from the person's him/herself. The inalienable characteristic of personal rights requires the prevention of gaining economic benefits from the human body parts by the rule of the prohibition of financial gain. As it is mentioned above, it is not possible for a person to sell his/her body parts for research and therefore, such research would be unpatentable. Human body parts of the donors- for our case of study their human embryos or HESCs- are protecting as being their personal data with their personal rights. The restriction of having financial gain from the person's own body parts is a necessity of human dignity.

The right to property is also claimed to be a tool for protecting departed human body parts. The right to property is one of the absolute rights. Absolute rights can be divided into two, such as; absolute rights on tangible properties and absolute rights on abstract entities. Absolute rights on tangible properties are rights that give its owner direct control over the property and such rights can be claimed against everyone. Absolute rights on the abstract entities are defined as intellectual property rights. The intellectual property rights are the rights of individuals regarding their thoughts, intelligence, knowledge, and feelings. They are the rights of the products in the field of art, literature, law, or science. Such a right to property is not constituted on the product, it is directed on the intellectual property which has a legal entity independent of the product itself. Both property rights on the tangible properties and the intellectual property rights provide a superior power of protection to its owner. In the HESCRs, the object of the patent application, the object that the right to property is wanted to be constituted on, is not the body part of the donor. The object of the patent application to which the protections of the right to patent is wanted to be granted is the method or the HESCL which is produced as a conclusion of the application of the method. So, it is not possible to the right to property is constituted on a body part. Besides, HESCs themselves is not patentable since such an object of the patent application would not have the criteria of novelty and inventiveness. The thing that makes HESCs patentable is the technic, the scientific method applied to it.

It is possible to mention intellectual property rights in the patentability of the methods of the HESCRs. However, in the patentability of the HESCLs, since the object of the patent application is the stem cell line which is produced as the consequence of the research, a property right on the tangible properties should be mentioned. While the right to the property gives several rights to its owner, it also imposes several obligations. In this scope, the kinds of embryonic stem cells that the right to property can be constituted, the techniques that can be patented, and the ways that the patents can be used should be evaluated under the obligations imposed by the right to property. One of the obligations imposed by the right to property and thus the patent is that a legal act contrary to ordre public and morality should not be protected by the legal system. For this reason, with this obligation imposed by the right to property, the patent owner cannot act against human dignity while exercising his/her rights [29].

To sum up, HESCs, which are a kind of body parts that can be departed from the body, can be protected with both personal rights and the right to property. The protection granted with personal rights is for the favor of donors. This protection appears by both taking their informed consent and prohibiting the financial gain. However, the right to property is for the favor of the owner of the invention. With the right to property of the owner of the invention, the technique he/she applies on the HESCs and/or the HESCLs produced as the consequence of the technique are protected. So, the object which provides financial gain and the right to property will be constituted on is the technique of the owner of the invention and/or the HESCLs produced as a consequence of the technique. Another value that should be considered is the principle of human dignity. It is possible to prevent the patent applications contrary to human dignity, with the restrictions provided by the personal rights and obligations imposed by the right to property. By giving the patent right for the HESCRs, this researches will have economic support, which will help the development of science. This way, there will be an opportunity to find and develop the treatments of severe human diseases.

4. Ethical/moral discussions on the patentability of the HESCLs

Ethical debates on both HESCRs and the patentability of the HECSLs find their basis in the debate on the legal/moral status of the embryo. We have mainly discussed the approaches to the legal status of the embryo in the first part of the study. The subject of legal status of the human embryo is mostly related to the ethical questions that arise due to the destruction of the embryo by HESCRs. In this part of the study, we will first consider ethical debates on stem cell discussions in general and then ethical debates on the patentability of the HESCLs based on these discussions.

The basis of the views which claim that HESCRs are unethical is the acceptance that the embryo is a part of human development. According to such views, embryo stage, which is a part of human development such as the stages of fetus, baby, child, old age, is also a humanbeing and is under the protection of human rights. Using the human embryo to treat diseases is the instrumentalization of humans. However, human should be an end, not a tool. This view may be called as "nongradualist position" [30].

It is clearly seen that the Kantian understanding of morality underlies this view. According to this, human beings are dignified and have certain rights due to their potential to be an intelligent and autonomous creature [31]. For the same reason, the embryo deserves the protection of human rights due to its potential to be a creature with intelligence and autonomy, just like other people who are lack of intelligence and autonomy. This view, in the discussion of Kantian moral philosophy, which will be discussed below, adopts the idea that the embryo does not have to have reason and autonomy, therefore, having this potential is sufficient to benefit from the protection of human rights [32].

The second argument in the debate about the status of the human embryo is that the human embryo is a creature to be protected and respected, but does not have the quality of a fully developed baby. The moral status of the embryo increases with its development. Once formed, it gains the right to be protected as a human and to have rights. In this mode of understanding, the moral status of the embryo is not absolute but related to other moral elements. So it is at a relative level. When it comes to the possibility and benefit of other people's treatment, the moral status of the embryo at a certain stage of development is decided by comparing it with this benefit. If the benefit to be achieved is a state of "goodness" that is superior to the destruction of the embryo, then destroying the embryo is not considered wrong. This argument provides an ethical opportunity for HESCRs [33].

Another question that needs to be answered after the question of whether the human embryo has the right to life is whether the human embryo is a carrier of human dignity. The main issue here is whether it is possible to talk about human dignity where there is no human life. For example, if we do not accept the embryo's right to life, will we be able to honor it? In the doctrine, it was argued that the right to life and human dignity should be evaluated separately. It has been determined that there is human dignity wherever there is human life, but it cannot be said that there is no human dignity where there is no human life. However, the reason why the human embryo is honored here is because the dignity of the potential future person is preserved [34].

In another view, which describes the relationship between human rights and human dignity in a similar way and is based on the idea that the right to life and human dignity can exist independently, a distinction is made between respect for human life and the right to life with regard to the status of surplus human embryos. The human embryo is respected in terms of respect for human life. The right to life, on the other hand, is evaluated independently of this respect and is recognized gradually depending on passing certain stages. The result of this is that in some cases the general well-being outweighs the respect for the life of the human embryo. However, according to this opinion, the fact that the embryo can be sacrificed for the good of the society does not mean that the human embryo is not honored. Even if the human embryo does not have the right to life, it has honor. This view is based on the assumption that the benefit of humanity is superior to respect for the life of the human embryos that will be destroyed under any circumstances and is criticized by its opponents as being consequential [35]. This view has also been criticized for its contradiction. It was stated that if the life of the embryo is respected, human rights should be given to it. It has been argued that the solution to this complicated issue would only be possible if the in vitro fertilization method is limited with fertilizing the OVAs that will be transferred to the uterus. Only in this case the human embryo is not instrumentalized and human dignity will be preserved [36], which is not possible for the current level of development of the assisted reproductive techniques.

John Harris, on the other hand, argues that the embryos that occur due to the in vitro fertilization method or miscarriages should be used instead of being wasted. He bases this thesis in the "Waste Prevention Principle". The other option, which is wasting the resources, argues that there are very strong moral reasons for using these resources for a useful purpose. He states that if the surplus embryos are already going to be destructed, it is not wrong to use them for a good purpose. He goes one step further and claims that organ transplantation from a fetus subjected to abortion is not different from organ and tissue transplantation from a cadaver [37].

Among the views which claim that HESCRs are supportable, there is another view that refers to the fact that the right to life is not an absolute right and that restrictions can be placed on this right in order to find treatment for incurable diseases. The basis of this view is the acceptance that the human embryo is a human and has the right to life. The view suggests that there is a conflict of interests between the right to life of the human embryo and patients seeking for treatment and in this conflict of interests finding treatment for severe diseases can be preferred [38].

The topic is getting a little more complicated in therapeutic cloning. Because, unlike the case of harvesting HESCs from surplus human embryos, in the therapeutic cloning the human embryos are created only for the purpose of harvesting HECSs. While the using surplus human embryos derived from the in assisted reproductive treatments is still discussive, it is claimed in the doctrine that creating embryos only to use them for therapeutic HESCRs or treatments of patients is much more problematic in the ethical aspect. In the initial phase of this technique, the nucleus removed from the patient is transferred to the OVA and a human embryo is created. The created human embryo however, is not transferred to the uterus and is used to obtain HESCs. Since the embryo has nearly the same genetic characteristics with the patient who will benefit from the HESCs, it is used in the treatment of the patient's diseases and in the process of organ formation for transplantation to the patient. The main problem in therapeutic cloning is that the human embryo is formed to be destroyed for the treatment of another person's diseases. In other words, the human embryo is produced to be destroyed for the benefit of another person or society.

According to the views that oppose to the therapeutic cloning technique, even if it has a therapeutic aim, this technique makes the embryo only a tool for the treatment of another person, and moreover, to know that the embryo will be destroyed at the very beginning of this process means the denial of any value of human life and accordingly the violation of human dignity [36]. As mentioned earlier, it has been suggested that the embryo can be attributed to pre-life dignity from the view of it has a potential to become a fully developed human being. However, in cloning for therapeutic purposes, the human embryo will not be able to benefit from honor protection retrospectively as the human embryo was not created to become a human being. In other words, since the human embryo does not have the characteristics of a fully developed human being, its dignity will not be violated [39]. In this context, we need to evaluate whether the embryo is dignified or not according to Kant's ethics, which is our mainstay of human dignity. First of all, it has been claimed that making a differentiation between human dignity and the right to life made in German Law in order to protect dignity in terms of embryos and the dead was incompatible with Kant's understanding of human dignity [40].

The view of the human embryo does not have human dignity is referred to by many authors by referencing the views of Kant. According to this, it is said that the embryo should not be honored as it does not have the characteristics of being self-aware, taking responsibility for its actions, acting independently, based on Kant's grounding of human dignity within the framework of moral autonomy. However, this view was opposed, and it was argued that Kant did not connect human dignity to actions and fulfillment of these acts. According to Kant, anyone who has the capacity to act morally should be considered as an honored person in an experimental manner. Every human being has the capacity to take moral action that comes from being human. In Kantian ethics, human dignity is a concept which belongs to the imaginable world (the universe of noumens). Since honor belongs to the imaginable world and has a super-experimental quality, criteria such as brain activity belongs to the world of phenomena (universe of phenomena), life capacity outside the mother's womb should not be taken into account. The embryo should be dignified from the moment of fertilization. According to this view, although the human embryo cannot use its capacity to act morally, it has been said that it should be regarded as an honored person because it has this ability [40, 41].

However, in the doctrine, there are views which also claim that it should be discussed briefly whether therapeutic cloning is an ethically problematic technique or not because finding treatments for severe human diseases is also a noble aim which serves to the human dignity. According to this view, in the therapeutic cloning, it is aimed to create a somatic cell or tissue type from another somatic cell. The fact that a step for this method is creating a human embryo and destroy it to harvest HESCs should not affect the whole purpose in a bad way. The created human embryo was a somatic cell which is processed to become another human cell, tissue or organ type, which means it was never going to be a fully developed human being. Indeed, the hope for finding and developing treatments for severe human diseases is an important and unignorable necessity. For this reason therapeutic cloning technique which is legally accepted in some states such as United Kingdom and Holland deserves to be discussed without pretending as if it is the same thing with reproductive cloning which aims to produce human beings [42].

When the human dignity is violated is another point that should be mentioned. Human dignity becomes meaningful as an absolute law, as the human beings see themselves as an end and treat all other people in the same way. It goes beyond trends, personal qualities and acquires a universal quality. Just as we cannot hold the person who is used as a tool legally responsible for a passive act carried out by destroying his will over another; likewise, we cannot hold people responsible for their actions against their wills. Conversely, man is responsible for what he does, not what he suffers. However, by acting contrary to the value of her/his own species, of which a she/he is a member, she/he can act inappropriate to her/his own human dignity. The prerequisite for being understood with human dignity and its being so valuable seems to be protected by a good understanding of this value and thus by considering it with this understanding and valuing it at least as much as the value it finds in its own person. In this context, any interference with the possible right to life of the embryo, regardless of whether it human dignity or not, can be regarded as a behavior that violates human dignity if it can be considered as a behavior contrary to the value of the species of which a human being is a member.

All these debates also constitute the basis for the debate on the patentability of the stem cell. According to our point of view, it would be appropriate to evaluate the ethical discussions about patentability of the HESCLs in the same way the discussions on HESCRs are evaluated. So it would be meaningful to divide the topic of patentability of the HESCLs into two such as the patent applications that use surplus human embryos for their methods and the patent applications that create human embryos for their methods.

The first criticism of the patenting of HESCLs is that the human embryo has been instrumentalized, made into something tradable and of commercial value, just like the HESCRs. However, as mentioned in the previous sections, the subject of the patent is not the human embryo or HESC itself, but the processed state of the HESCs taken from in. In other words, it is not possible to patent the human embryo.

This criticism should be considered separately in terms of the therapeutic cloning method, which includes the process of destroying the human embryo produced for the treatment of another person. As it will be remembered, this method was subject to heavy criticism that the human embryo, which has a life potential, would be instrumentalized in the production process, since human embryos were produced for the treatment of another person. If this view will be supported, the criticism that the human embryo is instrumentalized in the whole process in which it is already known that the human embryos will be used for therapeutic purposes and will be subject to patents after some procedures in which the stem cell in the human embryo is taken and subjected to a certain stage of this treatment would seem meaningful. When this point of view is accepted, developing a moral argument to meet the criticisms becomes striking.

On the other hand, considering the arguments for the ethical acceptability of HESCs, it becomes difficult to argue that patentability is a tool for instrumentalization. This will be the case especially for surplus human embryos or the human embryos with low result which are derived as a result of the application of assisted reproductive treatments. As stated above, within the framework of the principle that Harris named as the "Principle of Prevention of Waste", it is possible to say that the destruction of the human embryos that will already be destroyed for a good purpose such as treatment cannot be different from the organ transplantation from a patient with brain death. Here, the debate of whether instrumentalization by patenting would be possible will come to the fore, since it will not be possible to instrumentalize the human embryo with stem cell studies alone.

It will be better to emphasize once again that it would not be correct to consider the inventions to be made thanks to stem cells obtained by this method as if it is the same thing with the patenting of the human embryo. In our opinion, the human embryo is not suitable for patenting on its own. Its moral/legal status does not allow it. Just like organ donation, the human embryo should only be offered by donation, and it should be prohibited to sell it for financial gain. However, it is also clear that using surplus embryos for therapeutic purposes is a morally supportable way. For this reason, the use of surplus human embryos to find new treatment is morally acceptable, and should be encouraged and facilitated.

The fact that HESCRs are highly expensive should also be considered. The sustainability of these studies despite this cost depends only on the fact that some financial gains can be obtained thanks to the inventions that will emerge as a result of these studies. If this opportunity is not provided, these studies can be carried out in an extremely narrow and barren framework. By considering this point, it seems appropriate to provide the financial support required for the continuation of these researches and treatments, which serve a great purpose such as the benefit of humanity, through patents. However, since the issue of patenting stem cell studies is an issue that directly affects human life in many ways, just like organ transplantation, every stage should be handled with sensitivity and all kinds of abuse should be prevented. For this reason, a condition of patentability will be that the works to be carried out are in accordance with public order and ethical principles. In addition, prescribing appropriate restrictions, will prevent violation and abuse of rights.

5. Conclusion

As it is seen, misunderstanding the term of human embryonic stem cell line and pretending as if it is the same thing with human embryo that has a potential to become a human being causes it to be understood as an unpatentable value. Likewise, interpreting the right to property in its traditional definition causes biotechnological inventions to have difficulties to succeed the criteria of patentability. However, developments in the medicine and science, the need for treatment for the severe human diseases, and the high costs of the human embryonic stem cell researches make the patent protection a necessity for the human embryonic stem cell lines. So, modern interpretations for fundamental rights and principles and new regulations on the biotechnological inventions become a need for such researches to be made and researchers to be supported for finding and developing treatments.

For a view in the doctrine, which we are agree with, providing the protection of the right to property to the human embryos is possible and it does not mean that human embryo is in the low status of a property. The sui generis legal status of the human embryo needs a sui generis protection. According to our point of view, providing the protection of both the personal rights of the donor and the right to property of the owner of the invention does not reduce the legal/moral status of the human embryo. On the contrary, providing extra protection is a proof for its high moral/legal status. Thanks to the obligations the personal rights impose its owner, the donors cannot have financial gain from selling their human embryos. And thanks to the obligations the personal rights impose to the third parties, the researchers cannot use the human embryos without their informed consent. Otherwise sanctions of private and criminal law appears because of the violation of personal rights based on the protection of the personal data of the donors and the research and it consequences become unpatentable. The same situation also appears with the right to property and accordingly patentability. Thanks to the prohibiton of having financial gain from the human body and its parts, a researcher cannot patent the human embryo or human embryonic stem cell. The researcher has to find a novel, inventive, and industrially applicable method for producing human embryonic stem cell lines or invent a new method for a treatment. The owner of the patent is restricted with both the criteria of patentability and the principle of human dignity. By this way owner of the patent right is supported for making expensive and difficult inventions for the human diseases to be treated and has to do it appropriately to the personal rights of the donor, criteria of patentability and the principle of human dignity.

Author details

Tansu Sayar Kanyış*, Ezgi Arslan and Oğuzhan Kanyış
Faculty of Law, Izmir University of Economics, Izmir, Turkey

*Address all correspondence to: sayar.tansu@gmail.com

References

[1] Sadler, T.W, Langman's Medical Embryology, 13. ed. China: Wolters Kluwer; 2015. 400 p. DOI: 9781469897806

[2] Robertson A, Ethics and Policy in Embryonic Stem Cell Research. Kennedy Institute of Ethics Journal. 1999; 9(2): 109-136. DOI:10.1353/ken.1999.0013

[3] Hakeri H, Tıp Hukuku, 10. ed. Ankara: Seçkin Yayıncılık; 2017. 576 p. DOI: 9789750244544

[4] Özbilen, A. B, İnsan Kökenli Biyolojik Maddelere İlişkin Hukuki İşlemler, first ed. İstanbul: Vedat Kitapçılık, 2011. 370 p. DOI: 9786054446155

[5] Karakaya A. Kök Hücre Çalışmaları ve Etik: Türkiye'de İnsan Embriyosundan Elde Edilen Kök Hücreler Üzerinde Yapılan Çalışmalarda Etik Sorunlar, [thesis]. İstanbul: Fatih Sultan Mehmet Vakıf University; 2013.

[6] Counsil of Europe, Convention for the Protection of Human Rights and Dignity of the Human Being with regard to the Application of Biology and Medicine: Convention on Human Rights and Biomedicine [Internet]. 1997. Available from: https://www.coe.int/en/web/conventions/full-list/-/conventions/rms/090000168007cf98, [Accessed: 2020-09-06]

[7] Official Journal of the European Communities, Charter Of Fundamental Rights Of The European Union [Internet]. 2000. Available from: https://www.europarl.europa.eu/charter/pdf/text_en.pdf [Accessed: 2020-09-06]

[8] Turhan Esener T, Güven K. Eşya Hukuku. 8. ed. Ankara: Yetkin Yayınları; 2019. 676 p. DOI: 978-605-05-0491-0

[9] Hatemi H, Aybay A. Eşya Hukuku. 4. ed. İstanbul: Vedat Kitapçılık; 2014. 207 p. DOI: 9786054823116

[10] Sirmen L. Eşya Hukuku. 7. ed. Ankara: Yetkin Yayınları;2019. 778 p. DOI: 978-605-05-0481-1

[11] Ünal M, Başpınar V. Şekli Eşya Hukuku. 10. ed. Ankara: Savaş Yayınevi; 2018. 422 p. DOI: 9786057963345

[12] Aydın Ö. Üremeye Yardımcı Tedavi Yöntemlerinden Doğan Hukuki Sorumluluk. first ed. İstanbul: Legal Yayıncılık; 2014. 178 p. DOI: 9786054847242

[13] Norman M.F. When Did I Begin? Conception of the Individual in History, Philosophy and Science. first ed. New York: Cambridge University Press. 1988; 217 p. DOI: 10.1111/j.1467-8519.1989.tb00350.x

[14] Şehiral F, Patent Hakkının Korunması. first ed. Ankara: Turhan; 1998, 218 p. DOI: 9757425753

[15] Bak B, İlaçta Zorunlu Patent Lisansı. Ankara Barosu Dergisi. 2011:3:105-126. DOI:13009885

[16] Beyin S. Gen ve Kök Hücre Buluşlarının Patentlenebilirliği [thesis]. Ankara: Turkish Patent and Trademark Office; 2014.

[17] The European Parlament And The Council Of The European Union, Directive 98/44/EC Of The European Parliament And Of The Council of 6 July 1998 on the legal protection of biotechnological inventions [Internet]. 1998. Available from: https://eur-lex.europa.eu/legal-content/EN/TXT/?uri=celex%3A31998L0044 [Accessed: 2020-09-06]

[18] Kankanala, Complications in Patenting Biotech Inventions: A Peek At

US Law [Internet]. 2010. Available from: https://www.bananaip.com/ip-news-center/complications-in-patenting-biotech-inventions-a-peek-at-us-law/ [Accessed: 2020-09-06]

[19] United States District Court of Delaware, PharmaStem v Viacell Case US5,004,681-US5,192,553 [Internet]. 2004. Available from: https://www.casemine.com/judgement/us/5914b6b2add7b0493477a7d6 [Accessed: 2020-09-06]

[20] Nelson A: Obviousness or Inventive Step as Applied to Nucleic Acid Molecules: A Global Perspectiv. North Carolina Journal of Law&Technology. 2004: 1-14.

[21] Bhatti, Topic 3: Legal Requirements for Patentability and typical Parts of a Patent Application [Internet]. 2017. Available from: https://www.wipo.int/edocs/mdocs/mdocs/en/wipo_ip_cnx_17/wipo_ip_cnx_17_3.pdf [Accessed: 2020-09-06]

[22] The European Patent Convention, The 16th edition of the European Patent Convention [Internet]. 2016. Available from: https://www.epo.org/law-practice/legal-texts/epc.html [Accessed: 2020-09-06]

[23] United Nations, Convention on Biological Diversity [Internet]. 1992. Available from: https://www.cbd.int/doc/legal/cbd-en.pdf [Accessed: 2020-09-06]

[24] Uğurlu S. Bioethics and the Patent Eligibility of Human Embryonic Stem Cells-Related Inventions in Europe [Internet]. 2014. Available from: https://www.jstor.org/stable/j.ctv941w4v [Accessed: 2020-09-06]

[25] Oğuzman K, Seliçi Ö, Özdemir S, KişilerHukuku (Gerçekve Özel Kişiler), 18 ed. İstanbul: FilizKitabevi; 2019. 480 p. DOI: 9789753685412

[26] Sayar T. Legal Status of theUnbornandIts Protection Through Criminal Law [thesis]. Dokuz Eylül University Graduate School of Social Sciences Izmir: Department of Public Law Human Rights Law Program; 2018.

[27] Sayar Kanyış T, Human Embryonic Stem Cell Research: A Comparison Between Turkey and Iran. J Skin Stem Cell. Online ahead of Print; 6(2): e101509. doi: 10.5812/jssc.101509.

[28] Council of Europe, Convention for the Protection of Human Rights and Dignity of the Human Being with regard to the Application of Biology and Medicine: Convention on Human Rights and Biomedicine [Internet]. 1997. Available from: https://www.coe.int/en/web/conventions/full-list/-/conventions/rms/090000168007cf98 [Accessed: 2020-09-06]

[29] Tekinalp Ü. Kök Hücrenin Patentlenebilirliği [Internet]. 2016. Available from: https://bilimakademisi.org/wp-content/uploads/2016/04/K%C3%B6k-H%C3%BCcre.pdf [Accessed: 2020-09-06]

[30] Human Stem Cell Research, Report Committee on Culture, Science and Education, Rapporteur: Mr Wolfgang WODARG, Germany, Socialist Group, Avaiable from: http://www.assembly.coe.int/nw/xml/XRef/X2H-Xref-ViewHTML.asp?FileID=10259&lang=en [Accessed: 2020.09.06]

[31] Kant, Immanuel: Groundwork of the Metaphysics of Morals, Cambridge Texts in the History of Philosophy, Avaiable from: https://cpb-us-w2.wpmucdn.com/blog.nus.edu.sg/dist/c/1868/files/2012/12/Kant-Groundwork-ng0pby.pdf , [Accessed: 2020.09.06]

[32] Guenin, Louis M.: Morals and Primordials, Avaiable from:

https://science.sciencemag.org/content/292/5522/1659.full [Accessed: 2020.09.06]

[33] Aydın, Emre: Kök Hücre Çalışmalarında Etik, Hacettepe Tıp Dergisi 2005;36(4),198-202, Available from: http://www.deontoloji.hacettepe.edu.tr/ekler/doc/kok_hucre.doc , [Accessed: 2020.09.06]

[34] Rosenau, Henning: Yeniden Canlı Üretimi, Tedavi Edici Klonlama Tartışmaları ve Alman Kök Hücre Kanunu, Çev.: Hakan Hakeri: Tıp Ceza Hukuku 2003, 54-64. Akt. Nagehan Gürbüz, Biyo-Tıp Hukuku Bağlamında İnsan Onuru Kavramı Tartışmaları, Avaiable from: http://nek.istanbul.edu.tr:4444/ekos/TEZ/49752.pdf, [Accessed: 2020.09.07]

[35] Steering Committee on Bioethics, The Protection of the Human Embryo In Vitro, Report by the Working Party on the Protection of the Human Embryo and Fetus, Council of Europa, 2003. https://www.coe.int/t/dg3/healthbioethic/activities/04_human embryo_and_foetus_en/CDBI-CO-GT3(2003)13E.pdf , [Accessed: 2020, 09.06]

[36] Çoban, Aykut: Türkiye'de İnsan Embriyosu Üzerinde Araştırma Yapmanın Hukuki Sonuçları, Türkiye Barolar Birliği Dergisi, S: 86, 2009, pp. 210. Available from: http://tbbdergisi.barobirlik.org.tr/m2010-86-582, [Accessed: 2020.09.06]

[37] Harris, John: Stem Cells, Sex and Procreation, Cambridge Quarterly of Healthcare Ethics, Volume :12, Issue:4 2003, pp. 368-369. DOI: 10.1017/s096318010312405x.

[38] Erkan, Volkan "Kök Hücre Çalışmaları ve Etik", Felsefe Ekibi Dergisi, Y:2006, pp. 5. Avaiable from: http://www.felsefeekibi.com/dergi/s5_y21.html, [Accessed: 2020, 09.07]

[39] Rosenau, Henning: Avrupa Konseyi Biyo-Tıp Sözleşmesi'ne göre Embriyon Araştırmaları ve Tedavi Edici Kopyalama, Çev. Hakan Hakeri, KhukA, Kasım 2005, Y:8, pp. 141. Akt. Nagehan Gürbüz, Biyo-Tıp Hukuku Bağlamında İnsan Onuru Kavramı Tartışmaları, Avaiable from: http://nek.istanbul.edu.tr:4444/ekos/TEZ/49752.pdf, [Accessed: 2020.09.07]

[40] Rothaar, Markus: Human Dignity and Human Rights in Bioethics: the Kantian Approach, Medicine, Health Care and Philosophy, Volume:13, Issue:3, 2010, pp. 253. DOI: 10.1007/s11019-010-9249-0.

[41] Heper, Altan: Biyoetikte Hâkim Etik Teoriler, Hukuk Felsefesi ve Sosyolojisi Arkivi 20, pp. 119-120. Avaiable from: http://hfsa-sempozyum.com/wp-content/uploads/2019/02/HFSA20-Heper.pdf, [Accessed: 2020. 09.06]

[42] Katoğlu, T, European Convention on Human Rights and Biomedicine as a Part of Turkish Law. AÜHFD. 2006; 55: 157-193. DOI: 1301-130

Section 3

Therapies, *In Vitro* and Translational Research

Chapter 5

Combined Application Therapies of Stem Cells and Drugs in the Neurological Disorder Attenuation

Chia-Chi Chen, Ying-Ching Hung, Chia-Yu Lin, Hsiao-Yun Chen, Ping-Min Huang and Shao-Wen Hung

Abstract

Neurological disorders (NDs) are diseases of the central and peripheral nervous system that affected the hundreds of millions of people worldwide. Temporal lobe epilepsy (TLE) is a common NDs with hallucinations and disturbance of consciousness that cause the abnormal neurological activity in any part of brain. Neuroinflammation (NI) has been identified in epilepsy-related tissue from both experimental and clinical evidence and suspected to participate in the formation of neuronal cell death, reactive gliosis and neuroplastic changes in the hippocampus, may contribute to epileptogenesis. The NI is tightly regulated by microglia, but it is thought that excessive or chronic microglial activation can contribute to neurodegenerative processes. Therefore, the modulation of microglia responses may provide a therapeutic target for the treatment of severe or chronic NI conditions. Although the condition responds well to antiepileptic drugs (AEDs), there are still unresponsive to AEDs in about 1/3 of cases. Neural stem cells are the origin of various types of neural cells during embryonic development. Currently, many results of stem cell therapies in the animal experiments and clinical trials were demonstrated the efficacious therapeutic effects in the attenuated symptoms of ND. Therefore, the combined application therapies of stem cells and drugs may be a promising candidate for the therapeutic strategies of NDs, especially TLE.

Keywords: combined application therapies, stem cells, drugs, neurological disorder attenuation

1. Neurological disorders in people

When the disorders occurred in the brain and spinal cord included cranial nerves, peripheral nerves, nerve roots, autonomic nervous system, neuromuscular junction, and muscles, etc., these disorders were named neurological disorders (NDs). NDs are complicated and that are caused by a loss of neurons and glial cells in these injured areas in brain or spinal cord. NDs include epilepsy, Alzheimer's disease, Parkinson's disease (**Figure 1**), dementias, cerebrovascular diseases (stroke, migraine, and headache disorders, etc), multiple sclerosis, neuro-infected disorders, brain cancer-caused disorders, head trauma-caused traumatic nervous disorders, and malnutrition-induced NDs. Currently, NDs can affect about hundreds of millions of people worldwide. Among NDs, more than 50 million people have epilepsy worldwide [1].

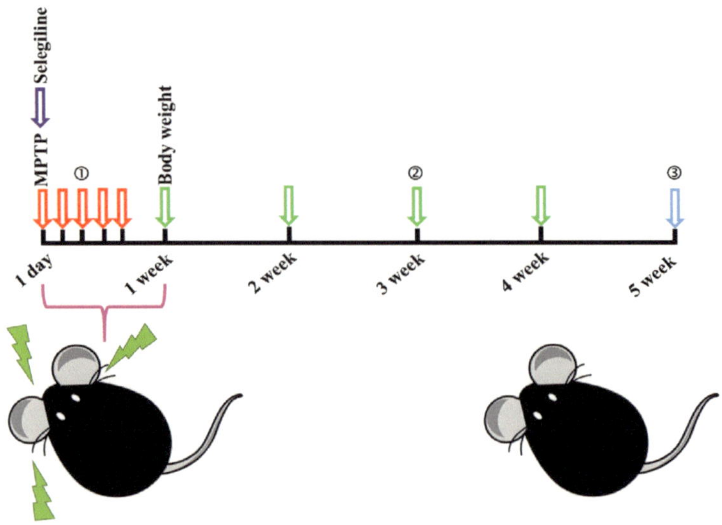

Figure 1.
Parkinson's disease model. Mice were treated with 1-methyl-4-phenyl-1,2,3,6-tetrahydropyridine (MPTP) by intraperitoneal injection once per day for continuous 5 days to induce Parkinson's disease ①. The experimental process of Parkinson's disease was described ②. ③ was the experimental end point.

2. Epilepsy

Approximately 1–3% of the worldwide population suffers from epilepsy. Pharmacotherapy represents the mainstay of treatment for most of the patient population. Moreover, surgery is another option for patients in whom a defined resectable seizure focus can be identified by brain imaging and seizure mapping techniques. For patients whose seizures cannot be controlled by antiepileptic drugs (AEDs) or not viable for surgery, vagal nerve stimulation will be a third possible option [2]. Unfortunately, there are a significant number of patients with epilepsy continue to live with uncontrolled seizures. Therefore, there is a clear need for more efficacious therapies for these epileptic patients with uncontrolled seizures. In order to identify an efficient therapy for preventing epileptogenesis, a validated animal model is required for translating discoveries at the molecular and genetic basis of epilepsy [3].

3. Neuroinflammation causes of epilepsy

Neuroinflammation is an underlying component of a diverse range of neurodegenerative diseases and their associated neuropathology, and increasing evidence suggests that microglia are a key causative factor in this process. Microglia comprised approximately 12% of cells in the brain [2]. They predominate in the gray matter especially in the hippocampus, olfactory telencephalon, basal ganglia, and substantia nigra. Commonly, microglia are readily activated in response to brain injury or immunological stimuli. Over-activated microglia can induce highly detrimental neurocytotoxic factors such as superoxide, nitric oxide (NO), and proinflammatory cytokines (IL-1β, IL-6, and TNF-α, etc.). In status epilepticus (SE), activated microglia have been shown to be present microgliosis [2, 3].
At present, many evidences have demonstrated rapid astrocyte and microglial activation following pilocarpine-induced seizures or SE. However, whether pilocarpine directly activate microglia is poorly understood.

4. Treatments of epilepsy

SE is a major neurologic emergency which is characterized by continuous seizures lasting greater than 5 minutes, and SE generation may increase the mortality and morbidity in patients [4–6]. The mechanism of SE generation is believed to correlate with the increase of excitatory activity and/or the decrease of inhibitory activity in the brain [7, 8]. Treatments can help epileptic patients to decrease seizures or stop seizures. These therapeutic treatments include medicines (antiepileptic drugs, AEDs), surgery, special diet (ketogenic diet), and electrical device therapy, etc. Traditionally, the strategies of SE treatment included the augmented inhibitory neurotransmission and/or reduced excitatory neurotransmission by pharmaceutical administration, but the clinically AEDs still fail to control the seizures in approximately 20–40% of SE patients [9–13]. Moreover, the underlying mechanisms of AEDs-resistant SE are still unclear, and it is important to figure out the response of pharmaco-resistance in intact brain for the development of promising treatment.

5. Characteristics, types, and therapy of stem cells

The treatment of epilepsy is not cured for a long time, and frequent seizures have brought endless troubles to the majority of patients. Nowadays, the treatment of epilepsy is no longer a medical problem, and can finally be completely ended. Only because of the advent of stem cell transplantation therapy, an indelible milestone has been set for the treatment of epilepsy. Stem cell transplantation technology has brought good news to patients with epilepsy, and has achieved a major new medical breakthrough in the treatment of epilepsy. Stem cell treatment of epilepsy is to inject healthy and young stem cells (SCs) into the patient's body through intravenous injection to replace their diseased or aging tissues and organs for the purpose of treating epilepsy.

Whether stem cell therapy effective for these epileptic patients who are not well treated by surgery? What is the principle? Stem cell transplantation increases the number of neurons In fact, it mainly uses the differentiation function of neural stem cells. Neural stem cells (NSCs) can divide and differentiate into corresponding cell types in the brain according to the induction of their surrounding microenvironment. Their morphology and function are very similar to those of nearby host cells, and they have a certain degree of safety. However, for drug-resistant intractable epilepsy, especially the effect of temporal lobe epilepsy (TLE) is better. TLE is caused by the degeneration, necrosis, and decrease in the number of hippocampal neurons, which leads to a decrease in the inhibitory neurotransmitter γ-aminobutyric acid. However, stem cell transplantation can repair and increase the number of γ-aminobutyric acid neurons. Therefore, restoring the balance of inhibitory neurotransmitters and excitatory neurotransmitters may fundamentally cure TLE.

The principle of cell therapy is easily described as your own cells as "autologous cells" or other people's cells as "allogeneic cells". After *in vitro* culture or processing procedures, these treated cells are introduced into the patient's body for apply to achieve the purpose of treating or preventing diseases [14].

SCs can build every tissue in the human body, there they have great potential for therapeutic applications in tissue regeneration and repair. SCs can differentiate into specific cell types. SCs' characteristics are the perpetual self-renewal and the differentiation ability to become a specialized adult cell type. Currently, two major classes of SCs are pluripotent SCs (PSCs) and multipotent SCs (MSCs) [14].

PSCs can differentiate into any cell in the adult body and it also was named "embryonic stem cells". Recently, "induced PSCs" (iPSCs), with some of the same PSCs' characteristics as proliferation, morphology and gene expression were yielded adult cells back into the pluripotent state via applying the molecular manipulation (added Klf4, c-Myc4, Lin28, and Nanog). Unfortunately, these iPSCs via genetic engineering are not likely approved for human disease therapy. Consequently, the results of a purely chemical approach to deliver safer transcription genes into the cells look promising [14].

MSCs are only restricted to differentiate a limited cell population. MSCs were found in bone marrow and have been used therapeutically since the 1960s. Recently, the new sources for MSCs such as the placenta, heart, umbilical cord blood, and brain were found. In the brain, neuro-progenitor cells have differentiated abilities to become neural cells. MSCs may have a potential for clinical application. MSCs have the plasticity to become all the progenitor cells or become only one or two specialized cell types of a particular tissue [14]. Based on our *in vivo* model, the results of SCs therapy in the epileptic research have some progress (**Figure 2**).

SCs can develop into cells of different specialized cell types in the body and grow into new body tissues. For example as transformation into different body cell types, including muscle cells, nerve cells, and blood cells. SCs have the potential to treat many diseases, including stroke, heart attack, spinal cord injury, and macular degeneration, etc. Whether SCs be used to treat epilepsy? Since most cases of epilepsy can be attributed to differences in receptor expression in the brain (due to mutations), theoretically correcting these may reduce the likelihood of electrical seizures in the brain. In addition, during status epilepticus, excitatory overload sometimes kills neurons, especially neurons in the hippocampus. Over time, this actually worsens the condition and leads to the development of temporal lobe epilepsy (TLE). Although AEDs can treat seizures, the damage to the temporal lobe is usually irreversible and permanent, and current treatments do not solve this problem [15–17].

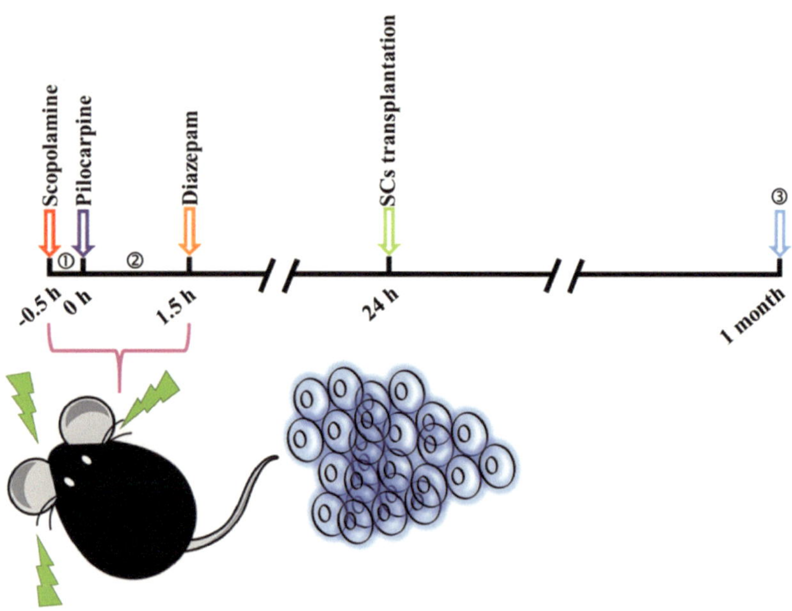

Figure 2.
Pilocarpine-induced status epilepticus model. Mice were firstly pretreated with scopolamine by intraperitoneal injection for 30 min before the induction of SE. Later, pilocarpine was administrated via intraperitoneal injection. Two hours after SE onset, mice were treated intraperitoneally with diazepam to terminate seizures. The process of status epilepticus was described as ① + ②. After one day post pilocarpine induction, stem cell therapy was performed until the experimental end point ③.

The scientific research of SCs in the treatment of epilepsy were showed as (1) PSCs transplantation to treat TLE: Human induced pluripotent stem cells (hiPSCs) are induced to develop into any type of cells in the body, including GABAergic interneurons. According to literatures, GABAergic interneuron progenitor cells derived from hiPSC were transplanted into the hippocampus of a TLE model to observe its efficacy. The results were presented that medial ganglion cells (MGE) derived from hiPSCs after the cell transplantation into the hippocampus. Later, researchers successfully reduced the frequency of seizures and reduced the loss of GABAergic neurons. (2) SCs were obtained from the patient's own bone marrow to treat TLE: the use of autologous mesenchymal stem cells in patients with epilepsy can reduce overall seizure frequency. SCs are obtained from the patient's own bone marrow and intravenously administered into the spinal cord. One year later, 30% epilepsy patients had a complete remission (without seizures), and 50% epilepsy patients who had not previously responded to the medication began to respond well [18–20].

6. Anti-epileptic candidate drugs

During neuro-inflammation, the pro-inflammatory cytokines as interleukin-1β (IL-1β), interleukin-6 (IL-6), and tumor necrosis factor-alpha (TNF-α) produced by activated microglia or astrocytes provoke pathological signaling cascades through the activation of phospholipase C and phospholipase A2. Later, the release of non-esterified arachidonic acid (AA) from phospholipids and the formation of lysophospholipids and bioactive eicosanoids by oxidized enzymes. Three enzyme systems, cyclooxygenases, lipoxygenases, and cytochrome P450 (CYP) epoxygenases metabolize released AA to lipid metabolites as prostaglandins, leukotrienes, and epoxyeicosatrienoic acids (EETs), respectively. Brain parenchymal tissue metabolizes AA to EETs via the CYP epoxygenase, which regulate cerebral blood flow and perform anti-inflammatory and anti-apoptotic effects. Previous results were showed that hypoxia and ischemic preconditioning experiments have shown that the increased expression of CYP epoxygenase and EETs in brain may confer protection from ischemic stroke induction. EETs signaling may play a role in suppressing the ischemia-evoked inflammatory cytokine responses in the brain circulation. Soluble epoxide hydrolase (sEH) is a key enzyme for metabolic conversion of EETs into their less active form, dihydroxyeicosatrienoic acids. The inhibition of sEH has been used to increase systemic EETs level and bioactivity. By using pharmacologic inhibitor or genetic deletion, the inhibition of sEH attenuated the vascular and neural injury induced by cerebral ischemia, suggesting sEH might be a novel target in treatment of stroke. In response to these findings, we hypothesized that sEH is involved in neuroinflammation-related epileptogenesis [21].

The sEH enzyme has thus been identified as a therapeutic target for inflammation, and therefore might serve to treat inflammatory pain [21]. Previous studies shown that during peripheral inflammatory, pain GABA A receptor mediated synaptic inhibition was enhanced in neurons [22]. Inhibition of sEH may have elevated the GABA agonists levels and enhance spinal GABAergic transmission [22]. In this context, the basis of kindling-induced epileptogenesis has been hypothesized that enhanced and/or attenuated activation in pathways utilizing glutamate and GABA as the neurotransmitter, respectively [22]. The sEH enzyme may influence the balance between neuronal excitation and inhibition, where glutamate and GABA, respectively, play important roles, and alter the seizure-induction threshold.

sEH is a member of the α/β-hydrolase fold family of enzymes which has two domains: a C-terminal epoxide hydrolase domain (EH), which is responsible for the biological roles associated with sEH and an N-terminal lipid phosphatase domain (PT) connected by a proline rich linker [23]. The role of sEH in lipid metabolism and lipid related disorders have been emerged in the recent years. sEH and dyslipidemia and related disorders such as atherosclerosis and coronary heart disease were related [24]. The products of the sEH hydrolase domain including fatty acid epoxide substrates and diol were found to activate PPARs, which can modulate plasma lipid by regulating lipid metabolism [24]. Moreover, the hydrolase domain of sEH are shown the effect on lowering cholesterol level by reducing the HMG-CoA reductase expression. However, the phosphatase domain of sEH was shown to regulate the cholesterol biosynthesis pathway by hydrolyzing the isoprenoid intermediates which are involved in cholesterol metabolism [24]. Summarily, the hydrolase and phosphatase domains of sEH exhibit opposite effects on expression of cholesterol levels. According the context about the opposite effects of sEH C- and N-terminal domain on regulating the cholesterol expression, we hypothesize that the sEH C-terminal domain may play a seizure-induction role and the N-terminal of sEH may be a seizure-protection role. In the previous studies, the pharmacokinetic parameters of sEH C-terminal inhibitor, AUDA [12-(3-adamantan-1-yl-ureido)-dodecanoic acid] in brain tissue samples were the mean residence time of 6 h and half-time elimination of 4 h [24].

7. Pilocarpine-induced SE models *in vivo* and *in vitro*

A pilocarpine-induced SE model is wildly used to investigate the alteration of neuronal circuits which reproduced most of the epileptic characteristics in patients. In the pilocarpine-induced SE model, acute SE is successfully induced in only a subset of rats after the high-dose pilocarpine injection [25]. According to our previous epileptic mouse model, mice were firstly pretreated with scopolamine methyl nitrate (1 mg/kg BW, Sigma-Aldrich) by intraperitoneal injection for 30 min before the induction of SE. Later, 325 mg/kg body weight (BW) pilocarpine hydrochloride was administrated via intraperitoneal injection. Mouse behaviors were monitored and scored according to Racine's scale (**Table 1**) [24–26]. Two hours after SE onset, mice were treated intraperitoneally with 10 mg/kg BW diazepam to terminate seizures (**Figure 3**).

In vivo seizure models are important for the research of epilepsy and seizure activity. Currently, *in vitro* models for epileptic research included rodent brain tissue slices and central nervous system (CNS) cell cultures [26]. These *in vitro* models have been performed for the research of epileptic mechanisms. The status of "seizure" and "seizure-like" were described in the *in vivo* and *in vitro* epileptic

Class (seizure stage)	Racine's scale behavioral expression
1	Mouth and focal movement
2	Head nodding
3	Contralateral forelimb clonus
4	Symmetrical forelimb clonus with rearing
5	Rearing and falling

Table 1.
Five class of Racine's scale for seizure stage.

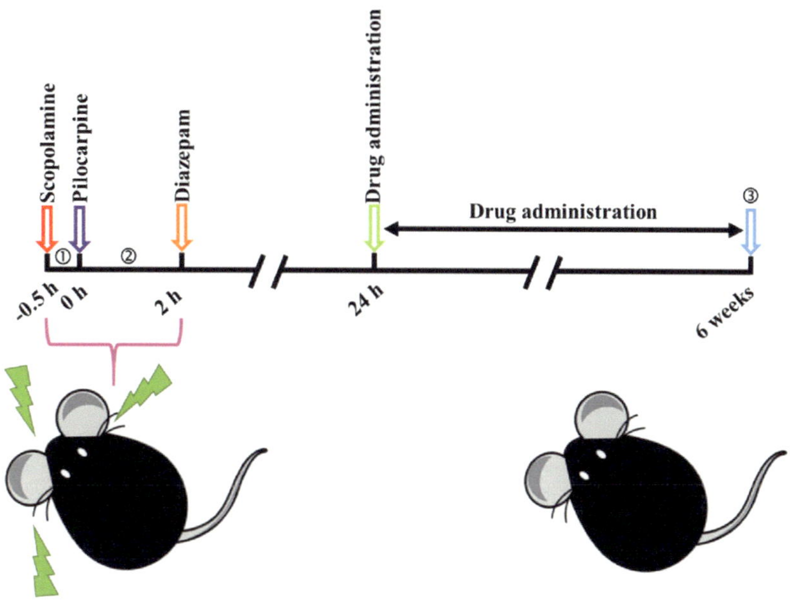

Figure 3.
Pilocarpine-induced status epilepticus model. Mice were firstly pretreated with scopolamine by intraperitoneal injection for 30 min before the induction of SE. Later, pilocarpine was administrated via intraperitoneal injection. Two hours after SE onset, mice were treated intraperitoneally with diazepam to terminate seizures. The process of status epilepticus was described as ① + ②. After one day post pilocarpine induction, drug therapy was performed until the experimental end point ③.

studies, respectively [27, 28]. *In vitro* CNS cell models were established by applying 12-(3-Adamantan-1-yl-ureido)-dodecanoic acid (AUDA) that was ordered from Cayman Chemical (Ann Arbor, MI, USA) and dissolved in dimethyl sulfoxide (DMSO; Cat No. 472301; Sigma-Aldrich, MO, USA). Cells were obtained as mouse retroviral immortalized microglia BV-2 cells belonged to C57BL/6 background (a gift from Dr. Hsiao-Li Chuang, National laboratory animal center, Taipei, Taiwan). They were grown in Dulbecco's modified Eagle's medium (DMEM) supplemented with 10% fetal bovine serum, 2 mM L-glutamine, 100 U/mL penicillin, and 100 μg/ml streptomycin in 5% CO_2 atmosphere at 37°C. Cells were grown to 90% confluency before experiments. Later, cells were treated with 100 μM pilocarpine and/or 100 μM AUDA and cultured for 24 hr. in 10% FBS-DMEM on glass coverslips and performed the observation of BV-2 cell morphology by light microscope. Cell viability was measured by MTT assay according to the manufacturer's instructions (MERCK, Darmstadt, Germany). *In vitro* scratch wounding-healing assay was performed as described. Briefly, confluent BV-2 microglia in 6-well plates were washed with serum-free DMEM three times. A line down the center of each well was scraped with a p200 pipette tip, followed by a wash to remove debris with serum-free DMEM. Images were taken at 10× magnification, scratch widths were measured, and wound closure was calculated by dividing widths measured after a 8 hr. incubation by the initial scraped width. Each experiment was carried out in triplicate and three fields were counted per well by scorers blinded to experimental conditions. Boyden chamber assays were performed in the transwells (BD Bioscience, New Bedford, MA) as previously described. BV-2 microglia (4×10^4 cells in 200 μL of serum-free DMEM) were added to the upper chamber and allowed to adhere to the polycarbonate filters (8 μm pore) for 30 min at 37°C in a humidified atmosphere of 95% air and 5% CO_2. Following, cells were pretreated with 100 μM pilocarpine at 37°C for 30 min prior to AUDA treatment. 100 μM AUDA were then placed in the upper chamber and lower chamber was added with 10% FBS-DMEM to allowed cell migration. Cells did not migrate and remained on the upper surface of the filter were removed. Cells had migrated to the lower surface were stained with Liu's stain (ASK, Taoyuan, Taiwan) and counted under light microscope. In at least three independent experiments, three wells per treatment were counted in nine random fields at 40× magnification per well by scorers blind to experimental conditions.

8. AUDA is safe in *in vivo* study and significantly inhibited pilocarpine-induced BV-2 microglial viability and migration *in vitro*

In the acute toxic assay *in vivo,* high doses of AUDA (100, 200, and 300 mg/kg BW) were respectively administrated for mice (n = 6/dose), all mice were survival and their clinical symptoms were normal. The results of AUDA were verified safe. Non-cytotoxic concentration (100 μM) of pilocarpine and AUDA were used in this study. Non-cytotoxic effect was presented after 100 μM pilocarpine combined with 100 μM AUDA treatment. According to our studies, AUDA significantly suppressed cell migration compared to the control by using respectively *in vitro* scratch wound-healing assay and Boyden chamber assay ($p < 0.001$) (**Figures 4** and **5**). According to these results, AUDA had the ability for the inhibition of the C-terminal domain of sEH can suppress the seizure development. C-terminal domain of sEH may play an important role in the epileptogenesis.

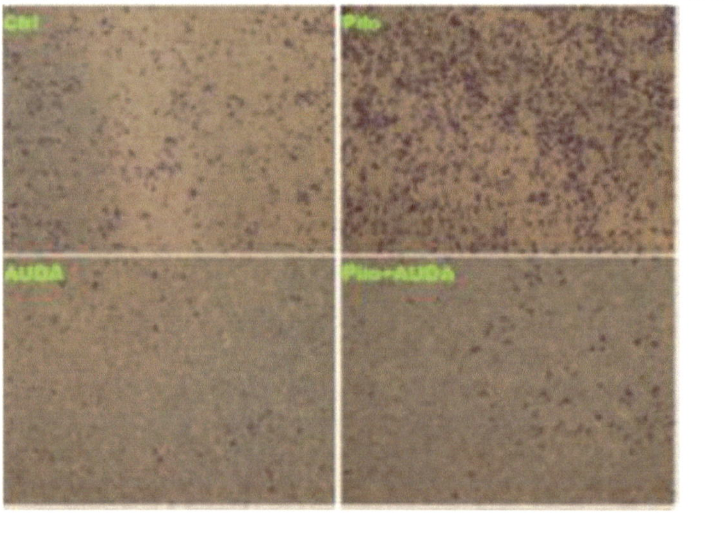

Figure 4.
In vitro *scratch wound-healing assay. Non-cytotoxic concentration (100 μM) of pilocarpine and AUDA were used in this study. 100 μM pilocarpine was first treated to BV-2 cells before 100 μM AUDA treatment. Later, AUDA treatment for 7.5 hours until the experimental end.*

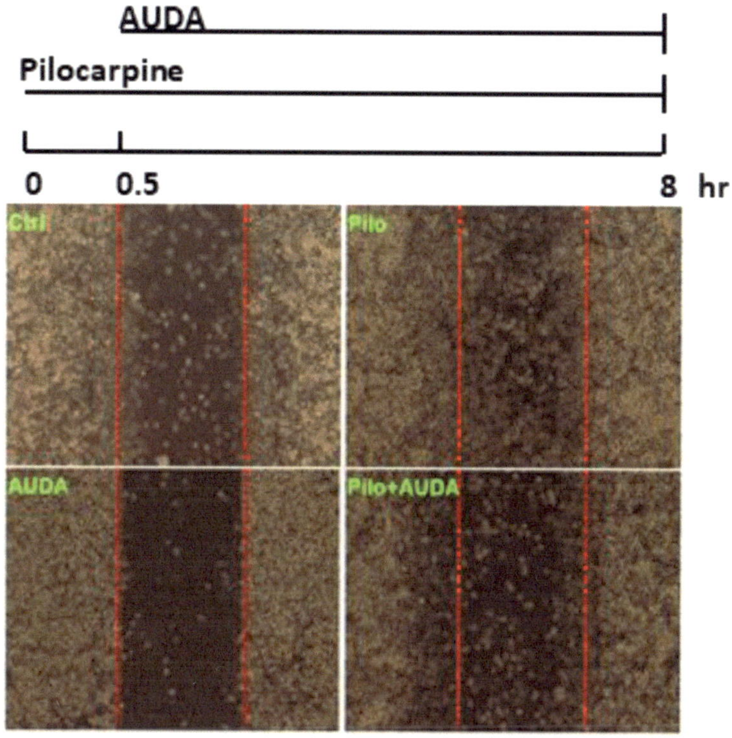

Figure 5.
Boyden chamber assay. Pilocarpine (100 μM) and AUDA (100 μM) were used in this study. Pilocarpine was first treated to BV-2 cells before AUDA treatment. Later, AUDA treatment for 7.5 hours until the experimental end.

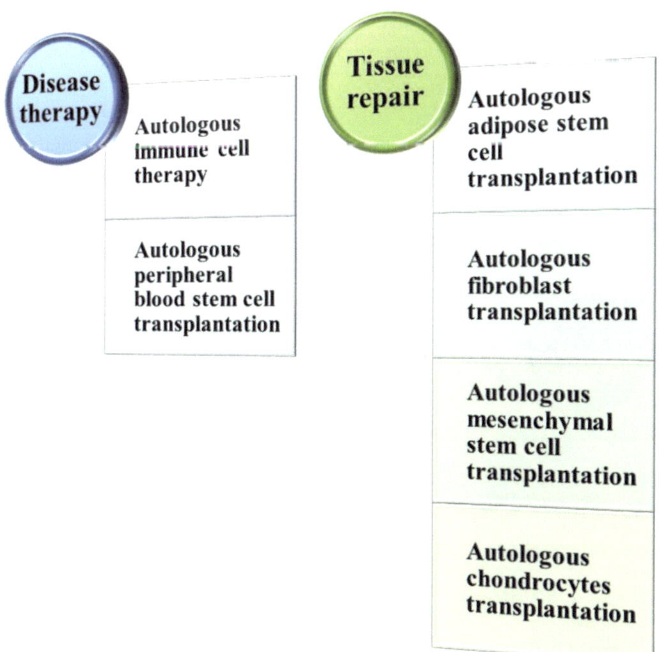

Figure 6.
Amendments to the administrative measures for the implementation or use of specific medical technology inspection and testing medical instruments in Taiwan in 2018. The six major projects can be divided into two categories: "Disease therapy" and "tissue repair". The two peripheral blood stem cells and immune cells are mostly used to treat cancer. The other four cell types belong to regenerative medicine, with the goal of helping tissue regeneration or repair.

9. Amendments to the administrative measures for the implementation or use of specific medical technology inspection and testing medical instruments in Taiwan

Cell replacement therapy and gene transfer to the diseased or injured brain have provided the basis for the development of potentially powerful new therapeutic strategies for a broad spectrum of human neurological diseases. However, the paucity of suitable cell types for cell replacement therapy in patients suffering from neurological disorders has hampered the development of this promising therapeutic approach. The full name of the "Special Control Law" issued by the Ministry of Health Services is "Administrative Measures for the Implementation or Use of Medical Instruments for Special Medical Technical Inspections", which was officially announced on September 6, 2018. The six major projects may seem complicated, but in terms of teleology, they can be divided into two categories: "disease therapy" and "tissue repair". The two peripheral blood stem cells and immune cells are mostly used to treat cancer. The other four cell types belong to regenerative medicine, with the goal of helping tissue regeneration or repair (**Figure 6**).

10. Conclusions

Currently, cell sources, characteristics, differentiation and therapeutic strategies and applications are frequently discussed worldwide. SCs have great potential in the regeneration and repair of tissue in people. However, limited information were presented at present. Therefore, the deeper studies will be still needs to be learned about their biology, manipulation and safety before their full therapeutic potential can be achieved. SCs-based therapies have shown encouraging results in treating diseases including epilepsy. Experimental animal and clinical studies have shown that SCs have significant regenerative properties on epileptic animal and patients with epilepsy. However, before SCs therapy becomes routine, larger clinical trials are needed [29–31]. In addition, our previous studies suggested that sEH played a

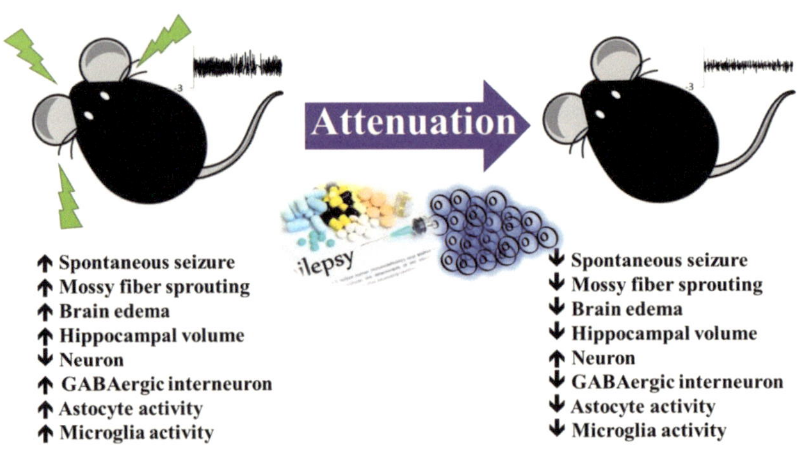

Figure 7.
Combined application therapies of stem cells and drugs in the neurological disorder attenuation. Based on epilepsy, some symptoms were increase as spontaneous seizure, mossy fiber sprouting, brain edema, hippocampal volume, GABAergic interneuron number, astrocyte activity, microglia activity. However, neuron number was decrease. After the combined application therapies of stem cells and drugs in the neurological disorder, the above symptoms were reversed.

critical role in regulating epileptogenesis in the pilocarpine-induced SE mice and *in vitro* CNS cell platform. Blocking of C-terminal activity of sEH via AUDA treatment can attenuate significantly epileptogenesis. Blocking of C-terminal activity of sEH on epileptogenesis may be provided a novel therapeutic approach in epilepsy. Anti-functional sEH C-terminal domain also may be a potential biomarker therapy for epileptogenesis in the future. Finally, the combined application therapies of SCs and drugs as AUDA may be try to apply in attenuating NDs (**Figure 7**).

Acknowledgements

Authors thank the Ministry of Science and Technology and the Council of Agriculture, Taiwan for support.

Conflict of interest

The authors declare no conflict of interest.

Author details

Chia-Chi Chen, Ying-Ching Hung, Chia-Yu Lin, Hsiao-Yun Chen, Ping-Min Huang and Shao-Wen Hung*
Aquatic Technology Laboratories, Agricultural Technology Research Institute, Hsinchu, Taiwan

*Address all correspondence to: 1032169@mail.atri.org.tw

References

[1] Vissers C, Ming GL, Song H. Nanoparticle technology and stem cell therapy team up against neurodegenerative disorders. Advanced Drug Delivery Reviews. 2019;**148**:239-251

[2] Hung YW, Lai MT, Tseng YJ, Chou CC, Lin YY. Monocyte chemoattractant protein-1 affects migration of hippocampal neural progenitors following status epilepticus in rats. Journal of Neuroinflammation. 2013;**10**:11

[3] Hung YW, Yang DI, Huang PY, Lee TS, Kuo TBJ, Yiu CH, et al. The duration of sustained convulsive seizures determines the pattern of hippocampal neurogenesis and the development of spontaneous epilepsy in rats. Epilepsy Research. 2012;**98**:206-215

[4] DeLorenzo RJ, Hauser WA, Towne AR, Boggs JG, Pellock JM, Penberthy L, et al. A prospective, population-based epidemiologic study of status epilepticus in Richmond, Virginia. Neurology. 1996;**46**:1029-1035

[5] Hanna J, Nichol A. Status epilepticus: an intensive care medicine problem. Anaesthesia & Intensive Care Medicine. 2012;**13**:148-151

[6] Loscher W. Molecular mechanisms of drug resistance in status epilepticus. Epilepsia 50 Suppl. 2009;**12**:19-21

[7] Bradford HF. Glutamate, GABA and epilepsy. Progress in Neurobiology. 1995;**47**:477-511

[8] Fritschy JM. Epilepsy, E/I balance and GABA(a) receptor plasticity. Frontiers Research Foundation. 2008;**1**:5

[9] Alldredge BK, Gelb AM, Isaacs SM, Corry MD, Allen F, Ulrich S, et al. Lowenstein DH. A comparison of lorazepam, diazepam, and placebo for the treatment of out-of-hospital status epilepticus. New England Journal of Medicine. 2001;**345**:631-637

[10] Alvarez V, Januel JM, Burnand B, Rossetti AO. Second-line status epilepticus treatment: Comparison of phenytoin, valproate, and levetiracetam. Epilepsia. 2011;**52**:1292-1296

[11] Claassen J, Hirsch LJ, Emerson RG, Mayer SA. Treatment of refractory status epilepticus with pentobarbital, propofol, or midazolam: A systematic review. Epilepsia. 2002;**43**:146-153

[12] French JA. Response to early AED therapy and its prognostic implications. Epilepsy Currents. 2002;**2**:69-71

[13] Wahab A, Albus K, Gabriel S, Heinemann U. In search of models of pharmacoresistant epilepsy. Epilepsia 51 Suppl. 2010;**3**:154-159

[14] Song CG, Zhang YZ, Wu HN, Cao XL, Guo CJ, Li YQ, et al. Stem cells: A promising candidate to treat neurological disorders. Neural Regeneration Research. 2018;**13**:1294-1304

[15] Wang D, Wang Y, Tian W, Pan J. Advances of tooth-derived stem cells in neural diseases treatments and nerve tissue regeneration. Cell Proliferation. 2019;**52**:e12572

[16] Luo L, He Y, Wang X, Key B, Lee BH, Li H, et al. Potential roles of dental pulp stem cells in neural regeneration and repair. Stem Cells International. 2018;**2018**:1731289

[17] Nguyen H, Zarriello S, Coats A, Nelson C, Kingsbury C, Gorsky A, et al. Stem cell therapy for neurological disorders: A focus on aging. Neurobiology of Disease. 2019;**126**:85-104

[18] Song CG, Zhang YZ, Wu HN, Cao XL, Guo CJ, Li YQ, et al. Stem cells: A promising candidate to treat neurological disorders. Neural Regeneration Research. 2018;**13**:1294-1304

[19] Angeloni C, Gatti M, Prata C, Hrelia S, Maraldi T. Role of mesenchymal stem cells in counteracting oxidative stress-related neurodegeneration. International Journal of Molecular Sciences. 2020;**21**:3299

[20] Hernández R, Jiménez-Luna C, Perales-Adán J, Perazzoli G, Melguizo C, Prados J. Differentiation of human mesenchymal stem cells towards neuronal lineage: Clinical trials in nervous system disorders. Biomolecules & Therapeutics. 2020;**28**:34-44

[21] Scharfman HE, Smith KL, Goodman JH, Sollas AL. Survival of dentate hilar mossy cells after pilocarpine-induced seizures and their synchronized burst discharges with area CA3 pyramidal cells. Neuroscience. 2001;**104**:741-759

[22] Isokawa M. Decrement of GABAA receptor-mediated inhibitory postsynaptic currents in dentate granule cells in epileptic hippocampus. Journal of Neurophysiology. 1996;**75**:1901-1908

[23] Wiebe SP, Staubli UV. Recognition memory correlates of hippocampal theta cells. Journal of Neuroscience. 2001;**21**:3955-3967

[24] Williams PA, Wuarin JP, Dou P, Ferraro DJ, Dudek FE. Reassessment of the effects of cycloheximide on mossy fiber sprouting and epileptogenesis in the pilocarpine model of temporal lobe epilepsy. Journal of Neurophysiology. 2002;**88**:2075-2087

[25] Hung YW, Hung SW, Wu YC, Wong LK, Lai MT, Shih YH, et al. Soluble epoxide hydrolase activity regulates inflammatory responses and seizure generation in two mouse models of temporal lobe epilepsy. Brain Behavior and Immunity. 2015;**43**:118-129

[26] Racine RJ. Modification of seizure activity by electrical stimulation. II. Motor seizure. Electroencephalogr Clin Neurophysiol. 1972;**32**:281-294

[27] Marchi N, Oby E, Batra A, Uva L, De Curtis M, Hernandez N, et al. *In vivo* and *in vitro* effects of pilocarpine: Relevance to ictogenesis. Epilepsia. 2007;**48**:1934-1946

[28] Wang CH, Hung CP, Chen MT, Shih YH, Lin YY. Hippocampal desynchronization of functional connectivity prior to the onset of status epilepticus in pilocarpine-treated rats. PLoS One 2012; 7: e39763.

[29] Shaimardanova AA, Solovyeva VV, Chulpanova DS, James V, Kitaeva KV, Rizvanov AA. Extracellular vesicles in the diagnosis and treatment of central nervous system diseases. Neural Regeneration Research. 2020;**15**:586-596

[30] Sadatpoor SO, Salehi Z, Rahban D, Salimi A. Manipulated mesenchymal stem cells applications in neurodegenerative diseases. International Journal of Stem Cells. 2020;**13**:24-45

[31] Wei WJ, Shi B, Guan X, Ma JY, Wang YC, Liu J. Mapping theme trends and knowledge structures for human neural stem cells: A quantitative and co-word biclustering analysis for the 2013-2018 period. Neural Regeneration Research. 2019;**14**:1823-1832

Chapter 6

Dental Stem Cell Banking and Applications of Dental Stem Cells for Regenerative Medicine

Karley Bates and Vincent S. Gallicchio

Abstract

Since the identification of mesenchymal stem cells, stem cell biology is a greatly researched field of regenerative medicine and tissue engineering therapies and has become an essential part of dentistry. Mesenchymal stem cells are multipotent stem cells that can differentiate into many cell types. Dental mesenchymal stem cell populations have been identified in dental pulp, human exfoliated deciduous teeth, periodontal ligament, dental follicle of third molars, tooth germ of third molars, gingiva of periodontium, alveolar bone, and apical papilla. Dental stem cells are the most natural, noninvasive source of stem cells that have been identified, and they have gained recent attention due to their accessibility and the associated relatively low cost of integration into regenerative therapy. Long-term preservation of dental stem cells is becoming a popular consideration and mirrors the ideology of banking umbilical cord blood. This review outlines the recent progress in the mesenchymal stem cells used in dentistry as well as some advancements that are being made in preserving dental stem cells for future personalized medicine. The aim of this study was to completely and concisely review the current use of adult dental stem cells specifically oral sources of stem cells, banking of dental stem cells, and applications or uses of dental stem cells specifically in oral regions and in a clinical setting.

Keywords: dental mesenchymal stem cells, stem cell banking, regenerative dentistry

1. Introduction

There exists an association between oral health and overall health throughout the lifetime of an individual. Because of this association, advancements are being made in dental practice to maintain teeth and orofacial bone as a way to combat general decline in oral health. Current common practice includes the use of artificial materials in dental implant procedures and other prosthodontic therapies. While often effective, these methods frequently require costly and invasive procedures and may result in failure or rejection. Recent research has given much attention to bettering therapeutic treatment and regenerative medicine associated with maintaining oral health. At the forefront of this research is investigating the use of stem cells in tissue regeneration, specifically in dental practices [1]. The use of stem cells in dentistry focuses on making the transition from using artificial materials in treatments to using individualized biological materials resulting in greater personalized medicine. The oral cavity has many locations from which to obtain stem cells, and stem cells can be harvested throughout many stages in an individual's life. Stem cell preservation through banking must be considered as a preventative way to obtain he biological materials necessary to make personalized medicine a common practice.

Adult stem cells, specifically mesenchymal stem cells (MSCs), can be loosely defined as immature multipotent cells that mature and multiply, and turn into many and more specific types of cells and tissues through the process of differentiation. The MSCs can be first derived from neural crest cells and pluripotent progenitor cells both of which develop during embryonic and child development [2]. A cell must be able to self-replicate and differentiate into two or more cell types to be considered MSCs [3]. Stem cells are the greatest source for means of regenerative medicine because they are autologous to the native tissue with low tumorigenesis risk and unlikely immune rejection [4]. Tooth derived MSCs, from both primary and permanent teeth, are of great interest regarding regenerative medicine because of how accessible they are due to the minimally invasive means through which these stem cells are obtained.

Stem cell use, specifically in dentistry, is on the forefront of research and clinical application because there are many sources of stem cells in the oral cavity. Treatment of a number of dental, oral, and craniofacial diseases using stem cells is currently being explored. Tooth-derived MSCs have great capability to proliferate and differentiate into many mesoderm lineages, and their performance aligns with MSCs from non-oral origins [5].

Discoveries are continuing to be made regarding which stem cells have the most advantageous characteristics for success in allogenic treatment in dental regenerative medicine. Dental stem cells come from a number of sources that can be obtained at different points throughout an individual's life. These sources include dental pulp [6], human exfoliated deciduous teeth (SHED) [7], periodontal ligament of periodontium [8], dental follicle of third molars [9], tooth germ progenitor cells of third molars [10], gingiva of periodontium [11], alveolar bone [12], and apical papilla [13–15]. These sources each have varying levels of proliferation and a range of differentiation capabilities.

Two types of stem cells are necessary for dental hard tissue regeneration: epithelial stem cells, responsible for the formation of enamel, and mesenchymal stem cells, responsible for the formation of dentin [16]. While research and clinical applications exist regarding the use of MSCs in dentistry, no information is currently available regarding dental epithelial stem cells [1, 16].

The success of stem cell therapy is exemplified through a process known as stem cell banking which includes harvesting, storing, and cryopreserving the cells. Stem cell banking highlights the ideology of individually personalized medicine being the best preparation for future treatment and is quickly increasing in research and clinical attention [17]. Advantages to dental stem cell banking include: (a) safety in harvesting, (b) low ethical concerns present compared to that of applications of embryonic stem cells [3], (c) noninvasive means of harvesting [7], and (d) autologous source of stem cells.

2. Background

MSCs are the group of stem cells that are the most heavily researched and commonly used in clinical application. Being multipotent, they possess the capability to mature and differentiate into multiple mesenchymal tissue lineages: mesodermal, ectodermal, and endodermal lineages [18]. MSCs are also characterized by their ability to self-renew and proliferate to maintain the source of undifferentiated stem cells. Additionally MSCs have been shown to possess immunomodulatory effects including angiogenesis, anti-inflammation, and antiapoptosis [1] showing their great potential for use in regenerative medicine. Furthermore, MSCs are associated with minimal ethical controversy and can be obtained from many adult tissues in the body [19]. The study of human MSCs began with their being identified and researched in bone marrow. Through experimentation, bone marrow derived stem cells (BM-MSCs) have been found to regenerate certain skeletal tissues including bone, cartilage, and adipose and fibrous tissues [20]. While research has shown effective use of BM-MSCs in tissue regeneration, the rather invasive measures through which these cells must be obtained make BM-MSCs an unlikely candidate for expansion and normalization of personalized medical treatments. Scientists since have investigated means to obtain and isolate MSCs from other various tissues in the body. Some of these include adipose tissue [21], endometrial tissue [22], salivary gland [1], umbilical cord [23], synovial membrane and fluid [24], blood [25], and oral sources [1]. MSCs from oral sources are among the most easily accessible stem cells and often may be obtained through minimally invasive measures. They have also been found to possess similar immunomodulatory characteristics as BM-MSCs [26]. Because of their accessibility, researchers continue to investigate their clinical applications in regenerative medicine. There are eight identified populations of MSCs found in teeth and their supporting structures of the oral cavity.

3. Dental pulp stem cells

In 2000 Gronthos *et al.* were the first to identify MSCs located in the dental pulp of teeth [6] that are today known as dental pulp stem cells (DPSC). These cells were investigated based on their supposed clonogenic and proliferative characteristics. Gronthos *et al.* suggested that DPSC have the progenitors required to amount similar cellular characteristics to those of BM-MSCs [6], a population of stem cells that is better understood. The research focused on comparing DPSC forming odontoblasts and BM-MSCs forming osteoblasts, specifically investigating if the DPSC possess the same proliferation and differentiation capabilities as BM-MSCs in both *in vitro* studies and *in vivo* by *ex vivo* transplantation into immunocompromised mice. Experimentation of DPSC mirrored the previously established biological pathways of BM-MSCs. Odontoblasts are responsible for the formation of dentin and are suspected to arise from the dental pulp, but little information is known about the specific cell population that gives rise to the odontoblasts in the dental pulp cavity [6] and what biological scaffold is most effective in aiding in the production of odontoblasts.

Gronthos *et al.* obtained pulp tissue from third molars of individuals aged 19–29. *In vitro* experimentation on differentiation potential of DPSC showed no formation of adipocytes, while BM-MSCs showed formation of adipocytes. Both cell populations were found to be clonogenic during *in vitro* manipulation. The DPSC were found to have a high proliferation capacity and proliferated more than the BM-MSCs [6]. BM-MSCs display bone formation on a hydroxyapatite/tricalcium phosphate (HA-TCP) powder, thus DPSC were observed in conjunction with HA-TCP powder to investigate if the DPSC would display a similar response as the BM-MSCs [6]. *In vivo* transplantation of DPSC resulted in odontoblast-like cells of human origin and contained the dentin-specific protein necessary to generate dentin [6]. This suggests the HA-TCP powder was conducive to the DPSC like the BM-MSCs.

Gronthos *et al.* concluded that the DPSC are able to be considered a source of stem cells because of their profound regeneration, differentiation, and proliferation capabilities. DPSC may be harvested from one pulp cavity and applied to dentin regeneration in many teeth [6]. Granthos *et al.* call for future research to investigate the biological scaffold that will amount to the most dentin regeneration associated with *in vivo* transplantation of DPSC.

4. Stem cells from human exfoliated deciduous teeth

In 2003 Miura *et al.* identified human exfoliated deciduous teeth (SHED) as a profound source of MSCs [7] to be considered for use in autologous regenerative therapies. The cells were researched using prior methodology of umbilical cord derived stem cells and investigated specifically regarding the prior knowledge and research pertaining to umbilical cord derived stem cells. The study focused on *ex vivo* success through *in vivo* transplantation. An advantage of SHED as a source of stem cells is that they are easily obtained from an individual at a relatively young age. SHED are found in the dental pulp that is left in deciduous teeth that are lost and replaced with adult permanent teeth during adolescence in an individual's lifetime. This makes SHED an ideal candidate for preservation through stem cell banking. The cells must be able to be preserved for use at a later time in the individual's life if they are to be considered a viable source of stem cells for regenerative medicine and treatment. Similar to DPSC, SHED provide a relatively large source of stem cells as cells harvested from one pulp cavity may be also used in regenerative treatment for many other teeth of the individual.

Miura *et al.* obtained pulp tissue from human exfoliated deciduous incisors from 7 to 8 year old individuals [7]. The researchers suggest SHED may be originating from a perivascular environment [7] due to their origin and association with vasculature in the dental pulp. The SHED were investigated in both *in vitro* studies and *in vivo* by *ex vivo* transplantation into tissues of immunocompromised mice. Single cells were isolated and allowed to proliferate to form populations of cells as a way to measure their characteristic clonogenicity. The study also showed that the SHED have greater *in vitro* proliferation capabilities regarding formation of cell populations as compared to both BM-MSCs and DPSC [7]. The cell populations were then induced to differentiate, and were successful in differentiating, into the following cell types *in vitro*: odontoblast-like cells, adipocytes, and neuronal cells [7]. Furthermore, the cells differentiated into odontoblast-like cells on HA-TCP powder and were able to survive after *in vivo* transplantation.

The same MSCs markers as BM-MSCs and DPSC have displayed were observed on SHED suggesting similar differentiation capabilities among the various sources of stem cells. However, SHED were found to differentiate into odontoblasts *in vivo* but were not successful in differentiating into dentin pulp-like complex like DPSC have been shown to do [7]. This suggests SHED possessing alternative differentiation capabilities or means of differentiating than those understood to be true of DPSC. Additionally, while the SHED were successful in producing odontoblast-like cells, SHED were unable to differentiate to produce osteoblasts. The cells were, however, able to induce varying degrees of bone formation in nearby host cells *in vivo* after transplantation. Miura *et al.* proposed the SHED may play a role in bone formation around permanent adult teeth as the deciduous teeth are exfoliated [7].

Miura *et al.* concluded that SHED are in fact a viable source of multipotent stem cells and have great potential in playing a role in regenerative medicine [7]. The researchers suggest the differences between SHED and DPSC may be attributed to the developmental differences between adolescent deciduous teeth that are meant to be exfoliated and the development of permanent adult teeth that are made to survive an individual's lifetime. Miura *et al.* call for future research to investigate the purpose of SHED in adolescent deciduous pulp cavities, as understanding their purpose may lead to greater understanding of SHED and thus greater application of SHED in regenerative medicine.

5. Stem cells from periodontal ligament of periodontium

The periodontal ligament (PDL) was found by Seo *et al.* to be a source of multipotent stem cells capable of playing a role in regenerating periodontal tissue [8]. The cells of the PDL were investigated based on the known differentiation properties of the PDL, and this information was then used to obtain evidence that the PDL obtains stem cells and may be used for generation of cementum and new PDL specifically *in vivo*. Additionally, cellular markers and other multipotent determining properties of periodontal ligament stem cells (PDLSC) were compared to those of both DPSC and BM-MSCs. There exist other populations of cells that are capable of regeneration and possess many osteoblast-like properties, but no prior evidence of a stem cell population had been observed in the PDL [8].

Seo *et al.* extracted third molars from 19 to 29 year old humans, and the cells from the PDL were isolated and induced to proliferate *in vitro* [8]. Interestingly, PDLSC were found to accumulate less calcium than DPSC [8]. Through *in vitro* studies, PDLSC were determined to have adipogenesis capabilities as well as proliferate into cementoblast and odontoblast-like cells. The *ex vivo* colonies were transplanted *in vivo* to immunocompromised mice through *in vivo* transplantation, specifically to the periodontal area of the mice. PDLSC developed cementum-like tissue different from tissues generated from BM-MSCs and DPSC suggesting a difference in differentiation capabilities of the three cell groups [8]. The PDLSC do, however, obtain the same cellular markers expressed by BM-MSCs and DPSC that suggest a potential perivascular origin [8]. The PDLSC were also able to interact physiologically with the host collagen fibers suggesting they possess the capability to differentiate into cells to produce collagen fibers as well [8]. The *in vivo* integration of PDLSC into the periodontal area of which they were transplanted suggests great potential for future application of periodontal tissue regeneration using PDLSC despite their failure to attach to the bone of the teeth [8].

Seo *et al.* conclude that PDL is a viable source for obtaining multipotent mesenchymal stem cells and is consistent with the clonogenic, proliferative, and differentiating characteristic of mesenchymal stem cells from oral sources [8]. This was also confirmed by Kawanabe *et al.* [27] and Menicanin *et al.* who confirmed the stem cell properties of PDLSC and further evaluated the post-transplant self-renewal potential of PDLSC [28]. Menicanin *et al.* further concluded that PDLSC possess significant capability to self-renew and regenerate periodontal tissue [28]. Onizuke *et al.* confirmed PDL as a source of stem cells and further analyzed the application of cell sheet engineer technologies in forming bioresorbable scaffolds to best support integration of the PDLSC for clinical applications [29]. Seo *et al.* call for future investigation of the cellular markers specific to PDLSC different from those expressed by BM-MSCs and DPSC. Additionally they conclude future research should analyze PDLSC and their PDL tissue regeneration on a larger quantity of test animals and a larger size test subject population [8].

6. Stem cells from dental follicle of third molars

Stem cells obtained from the dental follicle of third molars have also been determined to be a source of oral derived MSCs valuable in contributing to regenerative medicine [9]. The dental follicle was investigated based on its supposed capacity to differentiate into cementoblasts, osteoblasts, and periodontal ligament cells as well as its ease in accessibility from extracted teeth. Dental follicle cells possess the progenitors for osteoblasts that are responsible for aiding the development of many tooth related structures during growth and eruption of teeth and development of alveolar bone [9]. Findings from Gronthos *et al.* [5] research of DPSC were applied to the study of DFSC to identify if DPSC possess similar proliferation and differentiation capabilities.

Morsczeck *et al.* studied human dental follicle cells of extracted third molars [9]. The cells were isolated and induced to differentiate in both *in vitro* experimentation and *ex vivo* through *in vivo* transplantation using immunocompromised mice. A HA-TCP powder was included to induce uptake of the cells into the mice. It was found that DFSC behave similarly to the cells of the periodontium and are capable of regenerating cells types associated with the periodontium including alveolar bone, cementum, and periodontal ligament [9]. Morsczeck *et al.* concluded that DFMSC can play a role in bone tissue engineering and possess the necessary cellular markers to be considered stem cells [9].

A similar study was conducted by Park *et al*. that focused on comparing human skin derived MSCs, BM-MSCs, and DFSC [30]. Findings confirmed that DFSC have the capacity to differentiate into osteocytes, chondrocytes, and adipocytes *in vitro* [30]. Successful *in vivo* experimentation suggested that although the DFSC investigated possess MSC-like cellular markers similar to those of DPSC and BM-MSCs, further research should be conducted regarding the multipotency of the cells [30]. Osteogenic potential was however observed in both *in vitro* and *in vivo* cell populations of DFSC [30].

7. Tooth germ progenitor cells from third molars

In 2007 Ikeda *et al.* identified stem cells found in the tooth germ of developing third molars and termed them tooth germ progenitor cells [10]. The tooth germ includes many tooth-related tissues in the oral cavity, but more specifically refers to the developmental stage of these tissues [31]. The purpose of the study was to research the morphology, proliferation, and multipotency of the TGPC which were studied based on their ability to turn into hepatocytes mirroring the differentiation capabilities of BM-MSCs [32], but different from previously discussed orally-derived MSCs. The researchers focused on investigating TGPC specifically regarding their use in treatment of liver diseases with the goal of restoring liver function caused by liver disease through the use of stem cells as an alternative to drug therapy [10]. TGPC obtained at a young age from the tooth germ of developing third molars are associated with few ethical concerns due to their routine extraction that normally results in disposal making them a great source for potential harvesting of stem cells.

The cell populations were tested both *in vitro* and *ex vivo* through *in vivo* transplantation into immunocompromised mice with HA-TCP use in transplantation [10]. *In vitro* the cells were found to maintain their morphology and clonogenic properties through the isolation and cell population formation and showed the ability to differentiate into cells of three germ layers [10]. The cells possessed the cellular markers specific to mesenchymal cells, and transcription factors were observed suggesting pluripotency and confirming their stem cell characteristics [10]. The mesenchymal character of these cells was also confirmed by Yalvac *et al* in 2010 as well as their ability to differentiate into progenitor cells with adipocyte, neural, and osteogenic characteristics [31]. Yalvac and team also observed that TGPC expressed the cellular markers indicating stem cell character at a significantly higher level compared to DPSC [31]. *In vitro* cells displayed characteristics of osteoblasts, differentiated into neuronal cells and hepatocyte-like cells sharing characteristics of functional hepatocytes [10]. The Ikeda *et al.* were the first to show that the neural crest derived cells can differentiate into cells from the endoderm lineage [10]. After transplantation however, the *in vivo* cells showed great proliferation and indicated bone formation through the presence of osteocytes instead of only osteoblasts [10].

Ikeda *et al.* concluded that TGPC are able to play an important role in regenerative medicine specifically related to the development of hepatocytes. Successful cloning and proliferation in both *in vivo* and *in vitro* studies confirmed the stem cell qualities of TGPC on treatment of liver disease. The team calls for future research to be conducted investigating the capacity to which the application of TGPC differentiating into hepatocytes may be applied to regenerative medicine regarding other tissues [10], including tissues of the oral cavity.

8. Gingiva-derived mesenchymal stem cells

In 2009, Zhang *et al.* identified human gingival tissue as a source of MSCs to be considered for tissue regeneration and treatment [11]. The GMSC were investigated based on their abundance, accessibility, and tendency to be discarded as waste from surgical procedures. The focus of the study was to determine to what extent the GMSC are able to regenerate tissue and display associated anti-inflammatory characteristics, specifically in treatment of colitis [11].

The study was conducted using both *in vitro* and *in vivo* by *ex vivo* transplantation of human gingiva cells collected from the periodontium. The cells were isolated, then showed higher proliferation capacity compared to BM-MSCs [11]. *In vitro* the GMSC displayed successful differentiation into adipocytes and osteoblasts of the mesoderm lineage as well as endodermal and neural ectodermal cells suggesting stem cell character [11]. Additionally the GMSC were shown to possess the cellular markers specific to stem cells and phenotype associated with mesenchymal progenitor cells [11] further confirming their stem cell character.

The GMSC were transplanted using a HA-TCP carrier to mice infected with experimentally induced colitis. *In vivo* the cells were found to reduce disease severity and significantly reverse inflammation in the mice. Interestingly, the cells formed connective-like tissue instead of the expected osteoblast formation suggesting these cells may be useful in tissue regeneration [11]. Zhang *et al.* call for future research to investigate if osteogenesis may be induced *in vivo* with the aid of a different transplant carrier [11].

More recent research has shown GMSC ability to differentiate into osteogenic cells *in vivo* [33]. Yang *et al.* compared the capacity of both GMSC and PDLSC to periodontal disease treatment [33]. GMSC exhibit enhanced ability upon transplantation to combat inflammation compared to PDLSC regarding *in vivo* osteogenesis which is explained by the unique characteristics associated with the gingival tissue regarding its immunoregulatory function and expedited healing [33]. Yang *et al.* call for future research on determining the best feasible methods to incorporate the use of GMSC in therapeutic treatment of periodontal disease in tissue repair and regeneration in a clinical setting [33]. Tomar *et al.* compared GMSC and BM-MSCs specifically regarding if they may be considered a viable and sustainable source of stem cells to be used in regenerative medicine in a clinical setting [34]. GMSC compared to BM-MSCs were found to provide a more stable source of stem cells regarding ability to proliferate and retain morphology in a long-term culture setting as is necessary for clinical application [34]. This was also confirmed by Sun *et al*. [35] and Tomasello *et al*. [36] who agree GMSC should be considered as a source for generation of bone in regenerative medicine.

Continued research on GMSC examine the advantages the cells offer to cutaneous wound healing [37], tendon regeneration [38], as well as regeneration of peripheral nerve defects *in vivo* [39]. The wide array of applications and characteristics these stem cells possess deem them potentially useful in tissue regeneration and treatment specifically in the oral region and in a clinical setting.

9. Alveolar bone-derived mesenchymal stem cells

In 2005 Matsubara *et al.* identified the alveolar bone of the jaw as a viable source for stem cells to be used in autologous bone regeneration [12]. An advantage of ABMSC as a source of stem cells is that they are easily obtained from an individual during dental implant surgery or extraction of third molars as the alveolar bone is often exposed or accessible during both of these surgical procedures. ABMSC are often less painful to obtain compared to BM-MSCs from other bones in the body [12].

Matsubara *et al*. harvested human alveolar bone during normal third molar extractions [12]. The cells were researched using prior methodology of BM-MSCs and investigated specifically regarding the prior knowledge and research pertaining to BM-MSCs. Single cells were isolated and allowed to proliferate to form populations of cells as a way to measure their characteristic clonogenicity. The study focused on *ex vivo* success through *in vivo* transplantation into immunocompromised mice. The researchers compared the ABMSC to BM-MSCs from the iliac bone [12] regarding their proliferation and differentiation capacities. Upon examination *in vitro*, ABMSC were found to possess similar stem cell character and cellular markers as BM-MSCs regarding osteogenic proliferation, but displayed significantly less adipocyte and chondrocyte potential compared to BM-MSCs [12]. Furthermore, the cells differentiated into odontoblast-like cells on a calcium phosphate scaffold and were able to survive after *in vivo* transplantation. Additionally, an association between age of patient and proliferation was observed showing a decline in proliferation with increased age of individual from which the cells were obtained. This decline was noticeably larger than that of the MSCs from the iliac bone being studied. Matsubara *et al.* acknowledge the varying research regarding this association and suggest banking of the stem cells obtained from extracted third molars of individuals at a relatively young age suggests the greatest potential for application in a clinical setting [12].

Matsubara *et al*. concluded that ABMSC are in fact a viable source of multipotent stem cells and have great potential in playing a role in regenerative medicine, specifically regarding bone regeneration. They call for future investigation to determine if a different transport medium might increase the potential of ABMSC to differentiate into adipocytes and chondrocytes. Because of the strong osteogenesis potential of ABMSC, they are interested in determining the best clinical application of ABMSC in repairing bone in treatment of periodontal disease [12].

Another *in vivo* study focused on using stem cell sheets made from ABMSC as a non-scaffold approach for bone regeneration in the calvaria bone of rabbits. Liu *et al*. observed greater osteogenic differentiation in ABMSC than in long bone MSCs [40] showing greater mineral deposition and cellular markers indicating osteogenesis. However, similar research by Pettersson *et al.* showed that ABMSC and BM-MSCs display no significant difference in their ability to differentiate into osteogenic cells [41]. Research is unclear regarding comparing the differentiation ability of ABMSC and BM-MSCs, however research concludes that ABMSC are a potential source of MSCs to be considered for regenerative treatment in craniofacial defects in a clinical setting [40–42].

10. Stem cells from apical papilla

In 2006 Sonoyama *et al.* identified apical papilla from the root of human teeth as a novel source of mesenchymal stem cells to be considered for oral tissue regeneration [15]. The goal of the study was twofold: to confirm tooth apical papilla as a viable source of stem cells and to explore the combination of SCAP and PDLSC in regeneration of a root periodontal structure and associated tissues necessary to support an artificial crown [15]. A healthy and stable root is necessary to recover and maintain a functional tooth, and SCAP are thought to aid in regeneration of roots and surrounding tissues of teeth. SCAP provide a unique source of stem cells because they may be harvested from one source of apical papilla and applied to tissue regeneration in many surrounding teeth due to their great proliferation [15].

SCAP were collected and isolated from human apical papilla from normal extractions of third molars and allowed to proliferate to form cell populations for *in vitro* studies. The cells displayed cellular markers indicative of MSC character and were induced to successfully differentiate into osteoblasts and adipocytes [15]. Cells were then transplanted into immunocompromised mice with a HA-TCP carrier for further *in vivo* testing. The cells were able to generate both dentin and connective tissue in the *in vivo* model suggesting SCAP are a unique group of stem cells with different differentiation capacity than most dental derived stem cells [15]. In fact, SCAP showed greater differentiation into dentin than was expressed by DPSC [15].

An additional *in vivo* study was conducted using SCAP and PDLSC from swine followed by transplantation into immunocompromised mice. There exist similarities between human dental tissues and those found in swine thus conclusion can be made regarding human regenerative medicine based on the success of the tissue from swine [15]. Sonoyama *et al.* used SCAP and PSLSC from the swine and observed formation of cementum on the scaffold. Formation of both dentin and PDL are necessary to regenerate a viable root periodontal structure capable of supporting a tooth.

Sonoyama *et al.* propose that this is a viable method for functional restoration of teeth and supporting structures in a clinical setting [15]. They call for future research to be conducted regarding the immunogenicity of the cells and suggest that SCAP may be an ideal candidate for stem cell banking [15]. Current research focuses on identifying the precise biomaterials that make up the environment in which SCAP best contributes to tissue regeneration such as oxygen level [43] and ideal makeup of a transplant carrier [44, 45].

11. Current clinical applications

The current advancements and the recognition of the use of stem cells in regenerative medicine has resulted in human clinical trials. Various human clinical trials applying stem cells from different sources in treatment of oral conditions are presented below. Additionally, there continue to be preclinical studies aiming to perfect the transition from autologous source to implementation of stem cells prior to the application of research in human clinical trials. These preclinical studies often aim to determine which scaffolding methods and materials are the most promising to ensure effectiveness of stem cell application with minimal senescence [46]. The studies also attempt to identify the specific microenvironment conditions in which the cells are both taken from and placed for regenerative treatment and how it will impact the potential and effectiveness of the proliferation and differentiation capacities of the cells [47].

Nakashima *et al.* performed a pilot clinical study with five patients in which they analyzed the capacity of DPSC to regenerate dental pulp in patients with irreversible pulpitis [48]. They observed great success in all of their patients regarding regeneration of pulp tissue in the root canals as well as dentin formation in over half of the participants [48]. Although their endodontic study was successful, Nakashima and team call for larger clinical trials to be performed before implementation of their practices [48]. Xuan *et al.* performed a larger human clinical study with 20 patients expressing regeneration of pulp tissue to teeth that had experienced necrosis resulting from trauma [49]. The tissue regeneration included blood vessel and nerve regeneration and successful regeneration was still observed 24 months post stem cell implementation [49]. Raddall *et al.* further studied the specific scaffolding materials and methods for implementation of DPSC in endodontic regenerative treatment of the tooth root [50]. Their research both identifies successes and presents challenges with the scaffolds investigated. Their research, however, gives an example of how studies today are directed towards perfecting the methodology of implementation of stem cell therapies.

DPSC have also been successful in regenerating bone lost from periodontal disease [51–53]. Ferrarotti *et al.* observed a significant decrease in periodontal disease and an increase formation of bone in the intrabony defects when compared to the control patients of their study [51]. Hernández-monjaraz *et al.* observed decreased tooth mobility, a significant decrease in periodontal disease, and an increase in bone mineral density six months post operating [52]. Aimetti *et al.* also observed complete bone regeneration of intrabony defects on the mandibular bone in their study on a patient with chronic periodontitis [53]. All three studies conclude that the application of DPSC to bone periodontal disease shows significant improvement in a clinical setting with no observed adverse effects. Barbier *et al.*, however, observed no significant difference in mandibular post extraction socket healing resulting from use of DPSC [54]. Also in this study, bone density nor interdental septum height were found to be significantly increased by the addition of DPSC [54].

Additionally, PDLSC have been used to treat periodontal disease in the regeneration of periodontal tissues. In a recent study by Iwata *et al.*, bone regeneration decreased probing depth and increased radiographic bone depth in 10 patients six months after treatment [55]. The PDLSC were introduced using cell sheets into third molar extraction sites. Iwata *et al.* recognize PDLSC as a promising source for periodontal disease treatment and bone regeneration [55].

12. Discussion of stem cell banking

Stem cell banking may be loosely defined as the collection, isolation, and preservation of stem cells in preparation for use in regenerative treatment and therapy. The instance in which the cells are available for harvest might not align with when the cells are needed; thus the cells must be stored and preserved in a manner that will maintain the functionality of the cells over a potentially long period of time. The process is unique because it provides the means for individualized autologous medicine that can be implemented throughout an individual's lifetime [17]. Personalized advancements in medicine have the goal of increasing tolerance of the immune system and low rejection by host due to the host and the donor being the same individual. Specifically, dental stem cell banking utilizes the stem cells obtained from the oral region, and dental professionals, especially oral surgeons, are responsible for the collection of the cells. Dental stem cell banks were created in response to the identification of the oral region as a source of stem cells and current research investigates the best means of collecting, isolating, and preserving the cells so that the full potential of use of these cells may be reached in the advancement of personalized medicine.

The various oral sources of stem cells, as previously discussed, can be harvested during tooth extractions, the naturally occurring exfoliation of deciduous teeth, or through other minimally invasive surgeries. An advantage of utilizing orally derived stem cells is that there are fewer ethical concerns associated with their harvesting than with obtaining stem cells from embryological sources [17]. Additionally, the harvesting of DMSC is often simple and painless through a less invasive manner than stem cells harvested from other regions of the body as they are often a byproduct of a surgical procedure and would otherwise be discarded as waste. Dental stem cell banking, specifically using SHED, was found to be more cost efficient than banking stem cells obtained from cord blood [56]. If stem cells from cord blood were not obtained, stem cells from oral sources offer more opportunities in an individual's life to harvest and bank stem cells [56]. Stem cell banking is a proactive form of personalized medicine as the individual chooses to bank their cells in preparation for their future medical needs.

Current research regarding the banking of stem cells focuses on determining the specific conditions for optimal collection, isolation, and preservation of dental stem cells. By identifying the specific condition in which to harvest and store the cells, the proliferation and differentiation capacities of the stem cells can be maintained.

13. Collection

Stem cell banking begins in the dental chair with the collection of the cells during a scheduled extraction or routine eruption of the tooth. The dental professional assisting in the process works with the stem cell bank to decide regarding the specificities of harvesting the cell and initial processing steps. Dental stem cell banks currently exist in Japan, the United States, India, the United Kingdom, Germany, Singapore, Mexico, India and Norway. Dental stem cell bank information of FDA accredited banks in the United States is presented below:

USA:

BioEDEN, USA (https://www.bioeden.com/us/).

StemSave, USA (https://www.stemsave.com/).

Store-a-Tooth by Provia Laboratories, LLC, USA (http://www.store-a-tooth.com/).

National Dental Pulp Library, LLC (https://ndpl.net/).

Tooth Bank, USA (https://www.toothbank.com/).

It is important that the tooth from which the cells are collected is a normal healthy tooth with little to no decay due to pulpitis. Tsai *et al*. investigated the impact of dental pulpitis on the effectiveness of stem cell banking and found it to be ineffective overall [57]. A slower colonization process and increased proinflammatory cytokines were observed on the cells infected with pulpitis making then an unideal candidate for stem cell banking [57].

The cells are immediately placed in a container of sterile saline solution to maintain the vitality of the cells and to keep them from drying out during transport. The time between when the cells are collected and when the cells are processed is an important factor to determine the usefulness of the cells as the tissue may begin to degrade [58]. Perry *et al*. studied the impact of time between collection and processing and was able to obtain cells for culturing up to 120 hours after extraction [59]. This suggests that immediate processing post extraction is not a crucial factor necessary to effectively bank stem cells.

Perry *et al*. also investigated the most effective collection transport medium to be used in the transition from the dental office to the lab where the cells are isolated. No notable difference was found between using PBS compared to HTS. PBS is the cheapest and most accessible medium to store in a dental office, however, there exist other transport media that can be considered for use as well. Slight alterations in the numbers of cells that are obtained for processing from the different medias are negligible because as the cells are processed, enough cells will be proliferated in the growing cell cultures [59].

14. Isolation

Isolation refers to the processing of the cells in preparation for preservation by the stem cell bank that is preserving them. As previously mentioned, isolation of cells is possible up to 5 days after the tooth containing the cells is extracted or removed from its source [59]. The pulp from the harvested tooth is disinfected, and the cells are isolated and cultured in MSC media. The focus of current research is identifying the optimal conditions in which the cells are allowed to proliferate and colonize. Factors to be taken into consideration are the process through which the cells are isolated, the extent to which the cells are digested, the means of cellular attachment during culturing, and the media, serum, and supplements that are used to enhance cellular growth and function during the culturing process. Perry *et al.* identified proper sterilization of the harvested teeth as the most important aspect of forming uninfected and sterile cell cultures [59]. Additionally, they investigated the efficiency of the cells establishing cell cultures after being isolated from the harvested teeth. Cell colonies were present and displayed viable cells after 14 days of isolation [59]. Throughout the study, the stem cell populations were tested to confirm their MSC-like character. The cells possessed the necessary cellular markers characteristic of MSCs as well as displayed the ability to differentiate into osteogenic, adipogenic, and chondrogenic cells as is characteristic of MSCs [59].

15. Preservation - cryopreservation and magnetic field programmed freezing

Cryopreservation is the process of cooling cells to an eventual temperature below 150°C by liquid nitrogen vapor. The cells must freeze carefully and quickly as to not form ice crystals which would result in cell death [60]. This is the main cellular damage that a cell may experience as a result of cryopreservation as well as mechanical stress from ice formed outside of the cell [61]. A cryoprotective agent (CPA) can be used in the vitrification process to maintain cell functionality during preservation, but the cryoprotectants are often harmful to the cells or even can cause cell death [62]. Through cryopreservation, cells are able to maintain their cellular markers, gene expression, and differentiation potential [59]. Perry *et al.* determined this through the comparison of DPSC before and after cryopreservation [59]. However, some changes in proliferation capacity are observed before and after cryopreservation [63–65].

In 2010 Kaku *et al.* from Hiroshima University developed an alternative method for cryopreservation using a magnetic field programmed freezer instead of the established cryopreservation method. The novel method of preservation was created to combat both ice formation inside and outside of the cell as well as the toxicity associated with necessary levels of CPA [61]. This system of preservation is referred to as the cells alive system (CAS) and was adapted from an existing freezing methodology implemented in the food industry [61]. The CAS applies a weak magnetic field to the sample which lowers the freezing point. Kaku *et al.* were able to determine the optimal hold time and plunging temperature necessary to freeze the cells without inducing obstruction to the cellular membrane. Proliferation as well as differentiation capacity were observed for the cells to ensure they maintained their MSC-like character throughout the study [61]. Kaku *et al.* specifically investigated the magnetic field freezing using PDLSC and found that they could proliferate equally as the teeth that were observed immediately post extraction [61]. Another study by Lee *et al.* observed similar results upon investigation of DPSC compared to cells from teeth that were not preserved [66].

Interestingly, viable stem cells have not only been obtained from post preservation isolated cell populations but also from teeth that were preserved intact [59, 65, 67]. This confirms that immediate processing post extraction is not a crucial factor necessary to effectively bank stem cells which would in turn result in lowered costs associated with banking stem cells as the processing of the cells is the most costly part of the process [58, 59, 63].

16. Conclusion

With the discovery and research of dental stem cells come great opportunities to utilize these cells in regenerative medicine and dental tissue repair. The aspects of each source of stem cells must be analyzed so that the best candidate for stem cell banking and the specific regenerative treatment may be used in personalized medicine. The eight sources of stem cells and their ability to successfully aid in regeneration of tissues have been analyzed: human exfoliated deciduous teeth, dental pulp stem cells, periodontal ligament stem cells, dental follicle stem cells, tooth germ progenitor cells, gingival-derived mesenchymal stem cells, alveolar bone mesenchymal stem cells, and stem cells from the apical papilla. The relatively recent developments in stem cell biology make banking dental stem cells a feasible option for regeneration of tissues and other oral structures. As with any new developments in research, however, there exist certain limitations associated with achieving a consistent application in a clinical setting. Despite the many advancements that have been made in the field, the cellular conditions in which the stem cells may be collected, isolated, and preserved must be improved and perfected before clinical application is regularly implemented. The viability and stability of the preserved stem cells must be considered further so that the preservation efforts do not result in loss. Appropriate long term double-blind randomized clinical trials must be perfected before stem cell therapy may become a normalcy in a clinical setting. Additionally, current stem cell research is successful animal models but must be further applied to human models before clinical application may occur. The opportunity for immune rejection exists, but as is in the case with stem cells, immune rejection is limited due to the autologous nature of the cells. With continued research and developments in the field of stem cell biology, the dental stem cells support the continued development and betterment of the stem cell banking industry.

Abbreviations

MSCs	mesenchymal stem cells
SHED	stem cells from human exfoliated deciduous teeth
DPSC	dental pulp stem cells
PDL	periodontal ligament
PDLSC	periodontal ligament stem cells
DFSC	dental follicle stem cells
TGPC	tooth germ progenitor cells
GMSC	gingival-derived mesenchymal stem cells
ABMSC	alveolar bone mesenchymal stem cells
SCAP	stem cells from the apical papilla
BM-MSCs	bone marrow derived mesenchymal stem cells
DMSC	dental-derived mesenchymal stem cells
HA-TCP	hydroxyapatite/tricalcium phosphate

Author details

Karley Bates and Vincent S. Gallicchio*
Department of Biological Sciences, College of Science, Clemson University, Clemson, South Carolina, USA

*Address all correspondence to: vsgall@clemson.edu

References

[1] Paz AG, Maghaireh H, Mangano FG. Stem Cells in Dentistry: Types of Intra- and Extraoral Tissue-Derived Stem Cells and Clinical Applications. Stem Cells Int 2018; 2018.

[2] Caplan AI. Mesenchymal stem cells. J Orthop Res 1991;9(5):641-50.

[3] Mao JJ. Stem cells and the future of dental care. N Y State Dent J 2008;74(2):20-4.

[4] Ohkoshi S, Hirono H, Nakahara T, Ishikawa H. Dental pulp cell bank as a possible future source of individual hepatocytes. World J Hepatol 2018;10(10):702-7.

[5] Gronthos S, Brahim J, Li W, Fisher LW, Cherman N, Boyde A, et al. Stem cell properties of human dental pulp stem cells. J Dent Res 2002;81(8):531-5.

[6] Gronthos S, Mankani M, Brahim J, Robey PG, Shi S. Postnatal human dental pulp stem cells (DPSCs) in vitro and in vivo. Proc Natl Acad Sci U S A 2000;97(25):13625-30.

[7] Miura M, Gronthos S, Zhao M, Lu B, Fisher LW, Robey PG, et al. SHED: stem cells from human exfoliated deciduous teeth. Proc Natl Acad Sci U S A 2003;100(10):5807-12.

[8] Seo B, Miura M, Gronthos S, Bartold PM, Batouli S, Brahim J, et al. Investigation of multipotent postnatal stem cells from human periodontal ligament. Lancet 2004;364(9429):149-55.

[9] Morsczeck C, Götz W, Schierholz J, Zeilhofer F, Kühn U, Möhl C, et al. Isolation of precursor cells (PCs) from human dental follicle of wisdom teeth. Matrix Biol 2005;24(2):155-65.

[10] Ikeda E, Yagi K, Kojima M, Yagyuu T, Ohshima A, Sobajima S, et al. Multipotent cells from the human third molar: feasibility of cell-based therapy for liver disease. Differentiation 2008;76(5):495-505.

[11] Zhang Q, Shi S, Liu Y, Uyanne J, Shi Y, Shi S, et al. Mesenchymal stem cells derived from human gingiva are capable of immunomodulatory functions and ameliorate inflammation-related tissue destruction in experimental colitis. J Immunol 2009;183(12):7787-98.

[12] Matsubara T, Suardita K, Ishii M, Sugiyama M, Igarashi A, Oda R, et al. Alveolar bone marrow as a cell source for regenerative medicine: differences between alveolar and iliac bone marrow stromal cells. J Bone Miner Res 2005;20(3):399-409.

[13] Rai S, Kaur M, Kaur S. Applications of Stem Cells in Interdisciplinary Dentistry and Beyond: An Overview. Ann Med Health Sci Res 2013;3(2):245-54.

[14] Tamaki Y, Nakahara T, Ishikawa H, Sato S. In vitro analysis of mesenchymal stem cells derived from human teeth and bone marrow. Odontology 2013;101(2):121-32.

[15] Sonoyama W, Liu Y, Fang D, Yamaza T, Seo B, Zhang C, et al. Mesenchymal stem cell-mediated functional tooth regeneration in swine. PLoS ONE 2006;1:e79.

[16] Bluteau G, Luder HU, De Bari C, Mitsiadis TA. Stem cells for tooth engineering. Eur Cell Mater 2008;16:1-9.

[17] Arora V, Arora P, Munshi AK. Banking stem cells from human exfoliated deciduous teeth (SHED): saving for the future. J Clin Pediatr Dent 2009;33(4):289-94.

[18] Ullah I, Subbarao RB, Rho GJ. Human mesenchymal stem cells

- current trends and future prospective. Biosci Rep 2015;35(2).

[19] Xin WEI Xue YANG Zhi-peng HAN Fang-fang QU Li SHAO Yu-fang SHI. Mesenchymal stem cells: a new trend for cell therapy. 2013;34(6):747-54.

[20] Bianco P, Robey PG, Simmons PJ. Mesenchymal Stem Cells: Revisiting History, Concepts, and Assays. Cell Stem Cell 2008;2(4):313-9.

[21] Imamura H, Adachi T, Kin T, Ono S, Sakai Y, Adachi T, et al. An engineered cell sheet composed of human islets and human fibroblast, bone marrow-derived mesenchymal stem cells, or adipose-derived mesenchymal stem cells: An in vitro comparison study. Islets 2018;10(3):e1445948.

[22] Mutlu L, Hufnagel D, Taylor HS. The endometrium as a source of mesenchymal stem cells for regenerative medicine. Biol Reprod 2015;92(6):138.

[23] Ding D, Chang Y, Shyu W, Lin S. Human umbilical cord mesenchymal stem cells: a new era for stem cell therapy. Cell Transplant 2015;24(3):339-47.

[24] Harvanová D, Tóthová T, Sarišský M, Amrichová J, Rosocha J. Isolation and characterization of synovial mesenchymal stem cells. Folia Biol (Praha) 2011;57(3):119-24.

[25] Ab Kadir R, Zainal Ariffin SH, Megat Abdul Wahab R, Kermani S, Senafi S. Characterization of mononucleated human peripheral blood cells. ScientificWorldJournal 2012;2012:843843.

[26] Pierdomenico L, Bonsi L, Calvitti M, Rondelli D, Arpinati M, Chirumbolo G, et al. Multipotent mesenchymal stem cells with immunosuppressive activity can be easily isolated from dental pulp. Transplantation 2005;80(6):836-42.

[27] Kawanabe N, Murata S, Murakami K, Ishihara Y, Hayano S, Kurosaka H, et al. Isolation of multipotent stem cells in human periodontal ligament using stage-specific embryonic antigen-4. Differentiation 2010;79(2):74-83.

[28] Menicanin D, Mrozik KM, Wada N, Marino V, Shi S, Bartold PM, et al. Periodontal-ligament-derived stem cells exhibit the capacity for long-term survival, self-renewal, and regeneration of multiple tissue types in vivo. Stem Cells Dev 2014;23(9):1001-11.

[29] Onizuka S, Iwata T. Application of Periodontal Ligament-Derived Multipotent Mesenchymal Stromal Cell Sheets for Periodontal Regeneration. Int J Mol Sci 2019;20(11).

[30] Park B, Kang E, Byun J, Son M, Kim H, Hah Y, et al. In vitro and in vivo osteogenesis of human mesenchymal stem cells derived from skin, bone marrow and dental follicle tissues. Differentiation 2012;83(5):249-59.

[31] Yalvac ME, Ramazanoglu M, Rizvanov AA, Sahin F, Bayrak OF, Salli U, et al. Isolation and characterization of stem cells derived from human third molar tooth germs of young adults: implications in neo-vascularization, osteo-, adipo- and neurogenesis. Pharmacogenomics J 2010;10(2):105-13.

[32] Theise ND, Nimmakayalu M, Gardner R, Illei PB, Morgan G, Teperman L, et al. Liver from bone marrow in humans. Hepatology 2000;32(1):11-6.

[33] Yang H, Gao L, An Y, Hu C, Jin F, Zhou J, et al. Comparison of mesenchymal stem cells derived from gingival tissue and periodontal ligament in different incubation conditions. Biomaterials 2013;34(29):7033-47.

[34] Tomar GB, Srivastava RK, Gupta N, Barhanpurkar AP, Pote ST,

DOI: http://dx.doi.org/10.5772/intechopen.95407

Jhaveri HM, et al. Human gingiva-derived mesenchymal stem cells are superior to bone marrow-derived mesenchymal stem cells for cell therapy in regenerative medicine. Biochem Biophys Res Commun 2010;393(3):377-83.

[35] Sun Q, Nakata H, Yamamoto M, Kasugai S, Kuroda S. Comparison of gingiva-derived and bone marrow mesenchymal stem cells for osteogenesis. J Cell Mol Med 2019;23(11):7592-601.

[36] Tomasello L, Mauceri R, Coppola A, Pitrone M, Pizzo G, Campisi G, et al. Mesenchymal stem cells derived from inflamed dental pulpal and gingival tissue: a potential application for bone formation. Stem Cell Res Ther 2017;8(1):179.

[37] Zhang Q, Su W, Shi S, Wilder-Smith P, Xiang AP, Wong A, et al. Human gingiva-derived mesenchymal stem cells elicit polarization of m2 macrophages and enhance cutaneous wound healing. Stem Cells 2010;28(10):1856-68.

[38] Moshaverinia A, Xu X, Chen C, Ansari S, Zadeh HH, Snead ML, et al. Application of stem cells derived from the periodontal ligament or gingival tissue sources for tendon tissue regeneration. Biomaterials 2014;35(9):2642-50.

[39] Zhang Q, Nguyen PD, Shi S, Burrell JC, Cullen DK, Le AD. 3D bio-printed scaffold-free nerve constructs with human gingiva-derived mesenchymal stem cells promote rat facial nerve regeneration. Sci Rep 2018;8(1):6634.

[40] Liu Y, Wang H, Dou H, Tian B, Li L, Jin L, et al. Bone regeneration capacities of alveolar bone mesenchymal stem cells sheet in rabbit calvarial bone defect. J Tissue Eng 2020;11:2041731420930379.

[41] Pettersson LF, Kingham PJ, Wiberg M, Kelk P. In Vitro Osteogenic Differentiation of Human Mesenchymal Stem Cells from Jawbone Compared with Dental Tissue. Tissue Eng Regen Med 2017;14(6):763-74.

[42] Mason S, Tarle SA, Osibin W, Kinfu Y, Kaigler D. Standardization and safety of alveolar bone-derived stem cell isolation. J Dent Res 2014;93(1):55-61.

[43] Rémy M, Ferraro F, Le Salver P, Rey S, Genot E, Djavaheri-Mergny M, et al. Isolation and Culture of Human Stem Cells from Apical Papilla under Low Oxygen Concentration Highlight Original Properties. Cells 2019;8(12).

[44] Ha M, Athirasala A, Tahayeri A, Menezes PP, Bertassoni LE. Micropatterned hydrogels and cell alignment enhance the odontogenic potential of stem cells from apical papilla in-vitro. Dent Mater 2020;36(1):88-96.

[45] Tatic N, Rose, Felicity R. A. J., des Rieux A, White LJ. Stem cells from the dental apical papilla in extracellular matrix hydrogels mitigate inflammation of microglial cells. Sci Rep 2019;9(1):14015.

[46] Ma L, Hu J, Cao Y, Xie Y, Wang H, Fan Z, et al. Maintained Properties of Aged Dental Pulp Stem Cells for Superior Periodontal Tissue Regeneration. Aging Dis 2019;10(4):793-806.

[47] Zheng C, Chen J, Liu S, Jin Y. Stem cell-based bone and dental regeneration: a view of microenvironmental modulation. Int J Oral Sci 2019;11(3).

[48] Nakashima M, Iohara K, Murakami M, Nakamura H, Sato Y, Ariji Y, et al. Pulp regeneration by transplantation of dental pulp stem cells in pulpitis: a pilot clinical study. Stem Cell Res Ther 2017;8.

[49] Xuan K, Li B, Guo H, Sun W, Kou X, He X, et al. Deciduous autologous tooth stem cells regenerate dental pulp after implantation into injured teeth. Sci Transl Med 2018;10(455).

[50] Raddall G, Mello I, Leung BM. Biomaterials and Scaffold Design Strategies for Regenerative Endodontic Therapy. Front Bioeng Biotechnol 2019;7.

[51] Ferrarotti F, Romano F, Gamba MN, Quirico A, Giraudi M, Audagna M, et al. Human intrabony defect regeneration with micrografts containing dental pulp stem cells: A randomized controlled clinical trial. J Clin Periodontol 2018;45(7):841-50.

[52] Hernández-Monjaraz B, Santiago-Osorio E, Ledesma-Martínez E, Alcauter-Zavala A, Mendoza-Núñez VM. Retrieval of a periodontally compromised tooth by allogeneic grafting of mesenchymal stem cells from dental pulp: A case report. J Int Med Res 2018;46(7):2983-93.

[53] Aimetti M, Ferrarotti F, Gamba MN, Giraudi M, Romano F. Regenerative Treatment of Periodontal Intrabony Defects Using Autologous Dental Pulp Stem Cells: A 1-Year Follow-Up Case Series. Int J Periodontics Restorative Dent 2018;38(1):51-8.

[54] Barbier L, Ramos E, Mendiola J, Rodriguez O, Santamaria G, Santamaria J, et al. Autologous dental pulp mesenchymal stem cells for inferior third molar post-extraction socket healing: A split-mouth randomised clinical trial. Med Oral Patol Oral Cir Bucal 2018;23(4):e469-77.

[55] Iwata T, Yamato M, Washio K, Yoshida T, Tsumanuma Y, Yamada A, et al. Periodontal regeneration with autologous periodontal ligament-derived cell sheets – A safety and efficacy study in ten patients. Regen Ther 2018;9:38-44.

[56] Popuri SK. Concerns of a Pediatric Dentist in Dental Stem Cells: An Overview. Open Dent J 2018;12:596-604.

[57] Tsai AI, Hong H, Lin W, Fu J, Chang C, Wang I-, et al. Isolation of Mesenchymal Stem Cells from Human Deciduous Teeth Pulp. Biomed Res Int 2017;2017:2851906.

[58] Woods EJ, Benson JD, Agca Y, Critser JK. Fundamental cryobiology of reproductive cells and tissues. Cryobiology 2004;48(2):146-56.

[59] Perry BC, Zhou D, Wu X, Yang F, Byers MA, Chu T-G, et al. Collection, cryopreservation, and characterization of human dental pulp-derived mesenchymal stem cells for banking and clinical use. Tissue Eng Part C Methods 2008;14(2):149-56.

[60] Zhurova M, Woods EJ, Acker JP. Intracellular ice formation in confluent monolayers of human dental stem cells and membrane damage. Cryobiology 2010;61(1):133-41.

[61] Kaku M, Kamada H, Kawata T, Koseki H, Abedini S, Kojima S, et al. Cryopreservation of periodontal ligament cells with magnetic field for tooth banking. Cryobiology 2010;61(1):73-8.

[62] Fahy GM, Wowk B, Wu J, Phan J, Rasch C, Chang A, et al. Cryopreservation of organs by vitrification: perspectives and recent advances. Cryobiology 2004;48(2):157-78.

[63] Woods EJ, Perry BC, Hockema JJ, Larson L, Zhou D, Goebel WS. Optimized cryopreservation method for human dental pulp-derived stem cells and their tissues of origin for banking and clinical use. Cryobiology 2009;59(2):150-7.

[64] Ma L, Makino Y, Yamaza H, Akiyama K, Hoshino Y, Song G, et al.

Cryopreserved dental pulp tissues of exfoliated deciduous teeth is a feasible stem cell resource for regenerative medicine. PLoS ONE 2012;7(12):e51777.

[65] Lindemann D, Werle SB, Steffens D, Garcia-Godoy F, Pranke P, Casagrande L. Effects of cryopreservation on the characteristics of dental pulp stem cells of intact deciduous teeth. Arch Oral Biol 2014;59(9):970-6.

[66] Lee S, Chiang P, Tsai Y, Tsai S, Jeng J, Kawata T, et al. Effects of cryopreservation of intact teeth on the isolated dental pulp stem cells. J Endod 2010;36(8):1336-40.

[67] Gioventù S, Andriolo G, Bonino F, Frasca S, Lazzari L, Montelatici E, et al. A novel method for banking dental pulp stem cells. Transfus Apher Sci 2012;47(2):199-206.

Chapter 7

Induced Pluripotent Stem Cells from Animal Models: Applications on Translational Research

Laís Vicari de Figueiredo Pessôa, Naira Caroline Godoy Pieri, Kaiana Recchia and Fabiana Fernandes Bressan

Abstract

Over the history of humankind, knowledge acquisition regarding the human body, health, and the development of new biomedical techniques have run through some animal model at some level. The mouse model has been primarily used as the role model for a long time; however, it is severely hampered regarding its feasibility for translational outcomes, in particular, to preclinical and clinical studies. Herein we aim to discuss how induced pluripotent stem cells generated from non-human primates, pigs and dogs, all well-known as adequate large biomedical models, associated or not with gene editing tools, can be used as models on *in vivo* or *in vitro* translational research, specifically on regenerative medicine, drug screening, and stem cell therapy.

Keywords: pluripotency, regenerative medicine, stem cell, therapy, domestic animals, non-human primates

1. Introduction

For centuries, animal models have been used to aid on the quest for knowledge regarding human anatomy, physiology, and health, at first by simple observation, progressing to a proper investigation, selection of adequate models for given conditions and resuming on the development of specific transgenic animal models [1]. A recent concern regarding welfare and animal rights [2] has highlighted the relevance of in vitro models, such as pluripotent and adult stem cells. Here we describe the recent advances of biomedical research using induced pluripotent stem cell (iPSCs) models isolated from non-human primates, pigs, and dogs. Due to anatomical, physiologic, genetic, environmental, and other similarities to humans and conditions, those animals are considered highly relevant models for translational studies, each presenting specific advantages and drawbacks. Herein we discuss the advantages of using iPSCs, associated or not with gene editing tools, to enlarge the value and possible applications for pharmaceutical development and therapeutic approaches in these models.

 IntechOpen

2. Stem cells from animal models: applications on translational research

2.1 Non-human primates: most promising although challenging model?

Although non-human primates (NH-primates) represent only a small share of the animals used in medical research, the significance of those studies for human health, especially pharmaceuticals and new therapeutic approaches, is prominent [3, 4]. NH-primates are often the most suitable model for assessing the safety and efficiency of said drugs prior to human trials [5] and supply information to connect data from other relevant clinical models, such as rodents, to humans [6].

As study models, NH-primates are highly attractive due to longevity, behavioral, anatomical, genetic, physiological, and immunological similarities with humans [5, 7–10]. Over the past decades, NH-primates have been used on studies and research to prevent or cure human conditions, through the development of vaccines and drugs or treatment for cancer, diabetes, obesity, Parkinson's and other neurodegenerative, respiratory and cardiovascular diseases [4, 5, 11, 12], as well as methods to prevent mother-fetus transmission of diseases such as HIV [13], amongst other conditions and illnesses. Moreover, it has recently been shown that some NH-primates present a working memory capacity similar to that of human children [14], which highlights their importance for cognitive and neurological studies.

Stem cells are also considered an excellent tool for disease modeling and drug screening [15]. Although pluripotent cells derived from embryos, also called embryonic stem cells (ESCs), and multipotent adult stem cells (ASCs) are relevant and have been widely used on stem cell research and therapy purposes [16–23], ESCs limited sources and ASCs limited proliferation, and differentiation potentials have hampered their use. The advent of inducing pluripotency in vitro on virtually any somatic cell from any species reported since 2006, led to an entire flock of biotechnological and therapeutic applications. Thus, since the debut of induced pluripotent stem cells (iPSCs) [24], it is possible to produce patient-specific pluripotent stem cells that are highly valuable as models [15]. Furthermore, supported by age-related changes on the immune system of both humans and NH-primates [25], the use of said animals modeling human diseases associated with stem cell research might provide remarkable insight on translational stem cell-based therapy and transplantation [6].

Since iPSCs were first reported, these cells are now available for a variety of wild and domestic animal species (reviewed by [26]). Amongst NH-primates, they include but are not limited to the rhesus [27]; drill [28]; cynomolgus monkey [29, 30]; marmoset [31]; baboon [32]; orangutans [33]; Japanese macaque [34]. These cells were mainly generated from fibroblasts and integrative methods, but more recently, they were produced through non-integrative methods, such as Sendai-virus and episomal vectors [10, 35–37]. NH-primate-derived iPSCs have been used in research related to or as models for neurological [38–41], cardiac [36, 42, 43], reproductive [44], hematopoietic conditions [37, 45], transplantation and grafting [30, 46] and others.

As previously stated, similarities between humans and NH-primates make them essential models to assess the safety of drugs and therapeutic methodologies before human trials [5]. Immunologic similarities were considered when multiple NH-primate species were chosen as models to establish an iPSCs-derived multipotential hematopoietic progenitor cell differentiation protocol [37] and baboon enucleated red blood cells derived from iPSCs [45], aiming at blood disease and others preclinical testing. Cell transplantation is a relevant therapeutic methodology for some cardiac conditions leading to heart failure [42]. NH-primates iPSCs-derived cardiomyocytes were generated from rhesus monkeys [36, 43] and cynomolgus monkeys [42, 47] to assess drug screening, regenerative therapy, grafting viability, and immune rejection potential.

Aside from immunologic, physiologic, and genetic similarities, NH-primates cognitive capacity and longevity drag special attention for these animals as models for mental illness, age-related or not. Huntington's disease transgenic animals iPSCs have been used for generating neural progenitor cells that may be addressed for drug screening [48] pathogenesis modeling [40] and epigenetic and transcriptional profile analyses [49]. iPSCs and iPSCs-derived neural stem cells have also been generated from other NH-primate species aiming to develop regenerative therapy methods and modeling other neurological conditions, such as Alzheimer's and Parkinson's Disease [10, 36, 50–52].

Nevertheless, another possibility is the generation of custom-made specific transgenic disease models, by injecting retroviruses expressing target genes or gene editing techniques. The most known gene editing tools are zinc finger nuclease (ZFN), transcription activator-like effectors nuclease (TALEN), and clustered regularly interspaced short palindromic repeats (CRISPR). More recently, ZFNs and TALENs have been superseded by CRISPR/Cas9, which is equally, if not more efficient in inducing double-strand breaks (DSBs) and in stimulating homology-directed repair (HDR) [53, 54], also offering improved target specificity, prediction of off-target effects and activity [53, 55, 56]. Those approaches have been successfully applied to generate various NH-primate models (Reviewed by [57, 58]), including the above mentioned Huntington's disease transgenic monkey [59], Parkinson's [41, 60], neurodevelopmental disorders [61], Duchenne muscular dystrophy [62], severe combined immunodeficiency [63], and others.

Those models represent a significant scientific advance, allowing more faithful models than rodents previously used [58]. Although the use of NH-primate as research models is notable, some issues still need to be addressed. The greatly developed social skills of those animals implicate in environmental and social requirements to be met to keep NH-primates in an ethical and healthy environment ([64] art. 17), which implicates in high costs. Furthermore, results obtained from NH-primates studies are often not translatable to human research [65], highlighting the need for other research models, such as porcine and canine.

2.2 Swine: a large model in an already optimized production system

The domestication of swine (*Sus scrofa domesticus*) as a farm animal in established and controlled housing conditions, including specific conditions free of pathogens, has led to an important wide public acceptance that requires only minor adaptations for research [66]. The swine reproductive maturity is relatively fast compared to other large species (6–8 months), and they present a short gestation period (115 days) associated with the capability to produce large litters, with around 8–16 piglets per litter. Also, the swine body size, anatomy, physiology, and genetic homology are compatible with humans [66–69]. Hence, they are one of the most exciting species as a translational model for regenerative medicine research, and probably the most similar physiological model for humans apart from NH-primates.

The swine has already been explored as a biomedical model to develop diagnostic methods, studies, and treatment for several different conditions and diseases. For example, immunology studies and allergy models [70], and respiratory and cardiovascular conditions, such as pulmonary surfactant function, reperfusion injury, pulmonary hypertension, and asthma [71–73]. Similar to humans, swine are omnivores, reassuring its adequacy in studies examining the gastrointestinal system: transit time of pharmaceuticals [74], inflammatory bowel disease [75], gastric dilation [76] and metabolic disorders that influence of endocrine system [77–79]. The swine model has also been used to study neurological and neurodegenerative human disorders, such as amyotrophic lateral sclerosis, Alzheimer's Disease [80, 81], and Huntington's Disease [82].

For the advancement of regenerative medicine, specifically regarding cellular therapies, it is of great importance to study swine stem cells aiming to prove its efficacy and safety. Researchers have already demonstrated the effectiveness of treatment in the swine model using ASCs such as bone marrow mesenchymal stem cells (BM-MSCs) for the repair of myocardial infarction [83] and also for autologous therapy for disc degeneration [84].

However, cellular therapies using multipotent stem cells are restricted to specific diseases due to the limited capacity for differentiation to specific types of cells. Pluripotent stem cells, nevertheless, circumvents such drawback by presenting the ability to differentiate into several cells from the endoderm, mesoderm, or ectoderm origin, thus expanding the possibility translational studies for regenerative medicine [67].

ESCs are often studied and divided into two pluripotency states: naïve or primed. Naïve ESCs are found in the pre-implantation embryo, in the inner cell mass (ICM), and primed ESCs are found in the post-implantation stage in the epiblast [85, 86]. It is known that the mice ESCs cultured and maintained in vitro are considered "naïve", are collected from ICM and supplemented in culture with LIF, although human ECSs are collected from the epiblast and maintained in vitro with bFGF supplementation (for more details, refer to [87, 88]). For animal models including swine, the establishment of robust pluripotent ESCs using a straightforward and conventional approach has not yet been reported, and protocols regarding naive or primed pluripotency state characterization have not been consistent in the last decades [89].

Hence, the generation of iPSCs has shown to provide critical advantages over ESCs, particularly, when animal models are used. The iPSCs were already derived in the swine model (pig iPSCs or piPSCs) and reported in over 25 studies. The majority of those studies have used integrative methodologies to reprogram cells derived from embryonic, fetal, or adult fibroblasts. Although more efficient then non-integrative methods, integration of reprogramming factors onto the cell's genome might lead to the persistent expression of said factors, which can generate tumors and become unfavorable for cell therapy [90, 91]. Pluripotency induction using non-integrative vectors would greatly assist their use in cellular therapy [92]; however, piPSCs produced by episomal non-integrative methodology were until now only considered iPSCs-like [93].

piPSCs have already been induced to differentiate into several lineages: cardiomyocytes [94], hepatocytes [95], and even neuronal precursors cells [80, 96]. Kim et al. [96] for example, reported the derivation of piPSCs using porcine embryonic fibroblasts (PEFs) with four doxycycline-inducible human factors inserted into the cell by lentivirus, and the iPSCs generated were induced into neuronal progenitor cells (NPCs), positive for neuronal cells markers (PLAG1, NESTIN, and VIMENTIN). The differentiation protocol of iPSCs into NPCs can assist in future studies on animal models for neurodegenerative diseases, and the transplantation of these cells may provide details regarding the regenerative potency in vivo.

In particular, the swine is an attractive model to study human genetic diseases due to the genetic homology found between the species [97–99]. The extension of genetic editing tools to the piPSCs could significantly increase their value as a biomedical model, motivating efforts to develop safe and efficient genome editing technologies in this model, aiming to replicate human disease and develop therapeutic approaches [100].

In swine, gene editing tools are more prone to be effective and accepted once reproductive biotechnologies (such as embryo manipulation and microinjection and somatic cell nuclear cloning – SCNT) are far more studied than other models such as NH-primates and dogs [101]. The use of CRISPR/Cas9 injection into swine zygotes, for example, has been reported as an exciting model for human disease based on gene knock-out [102–104], in special, presenting high efficiency and without detection of off-targets [105].

Gene editing is highly explored in human iPSCs for cardiovascular, neurodegenerative diseases like Alzheimer's and Parkinson's, and degenerative muscular dystrophy (DMD), however, its applicability in autologous therapies is still restricted. Thus, gene editing in piPSCs to study diseases and their treatments [106] and transplant these cells or even to generate new entire edited organisms is a game-changer in the regenerative medicine field. Yu [104] edited swine zygotes using CRISPR/Cas9 for DMD the piglets born had the disease in skeletal muscle, heart and decreased smooth muscle thickness in the stomach and intestine. These models would enable, trough gene editing on piPSCs, to test autologous therapies for DMD.

Apart from the use of edited cells for cellular therapy, the technology would also be useful to the production of human organs by interspecies blastocyst-iPSCs complementation [68, 107]. Wu [108, 109] described the chimeras' production through the complementation of hiPSCs in swine zygotes genetically edited via CRISPR/Cas9. Researchers also reported to efficiently disable pancreatogenesis in pig embryos via zygotic co-delivery of Cas9 mRNA and dual sgRNAs targeting the PDX1 gene. When combined with chimeric-competent human pluripotent stem cells, the authors inferred that these results would provide a suitable platform for the xeno-generation of human tissues and organs in pigs [108, 109].

Bypassing the ethical problems of possible humanization of the swine during the embryo complementation process, another option for producing patient-specific organs is to recellularize swine organ scaffolds with hiPSCs. The selected organ goes through the decellularization process that completely removes cells and organic components of tissue, such as lipids, DNA, and antigenic proteins, but maintains the extracellular matrix (ECM). Recently, Goldfracht [110] combined hiPSCS-cardiomyocytes (hiPSC-CMs) with extracellular-matrix (ECM) derived from decellularized swine hearts, developing an ECM-derived engineered heart tissues (ECM-EHTs) model. Ohata and Ott [111] decellularized the lungs of human, swine, and NH-primates, the structure kept the original bronchial tree, vascular network, and most of the ECM composition and bioreactors were used to recellularize the lungs, and successful cell growth was achieved with perfusion culture.

Organ engineering based on recellularization with patient-derived iPSCs offers the unique potential to promote autologous treatment, and are also promising as tools for animal production, once piPSCs differentiated into germinative cells could be used to re-colonize depleted ovaries or testicles in order to spread the desired genetics in other animals [112]. Also, the use of muscle differentiation from iPSCs in scaffolds would benefit not only the cellular therapy for injured muscles in general, but opens new possibilities regarding in vitro meat production. The production of "animal-free" meat offers a reduction in environmental pollution and allows disease-free meat production due to its controllable and manipulative production system. However, technical challenges and intense research are still needed to establish such "animal-free" meat culture system [68, 113].

Although the complete reprogramming of iPSCs in the swine model is not yet fully elucidated as it is for human and murine reprogramming, the technology has the clear benefit of improving animal production and reproduction, opening new perspectives to study genetic diseases or develop cellular transplantation therapies.

Studies are still needed to optimize the production of non-transgenic piPSCs and their association with other biotechnologies in the swine model, and the inter-species difference regarding pluripotency acquisition is important in order to define proper culture conditions to maintain the pluripotency and the reprogramming protocols in this model.

2.3 Canines: closer to humans than ever

The dog (*Canis lupus familiaris*) is considered a well-suited animal model for many diseases, drug development, and regenerative therapies. Like humans, dogs present a great phenotypic diversity and a well-mixed gene pool because of centuries of random breeding [114], and they also exhibit metabolic, physiological, and anatomical similarities to humans [115]. More than 200 known hereditary canine diseases have an equivalent human disease, including cardiomyopathies, muscular dystrophy, and cancer. Moreover, the dog was the most prevalently used species in early transplantation research, including bone marrow transplantation and gene therapy [116], due to their similarities to humans concerning stem cell kinetics, hematopoietic demand, and responsiveness to cytokines [117, 118].

Because of the many similar cancer characteristics in dogs and humans, including histological features, genetics, behavior, and response to conventional therapies [119], dogs are amongst the leading models for human cancer studies. Notably, the number of dogs that are diagnosed and managed with cancer is estimated to be over 6 million per year in the United States [119]. Such conditions triggered researchers' interest and efforts to identify cancer-associated genes, study the environmental risk factors, understand tumor biology and progression, and develop of novel cancer therapeutics [120]. Different researchers described similar types of cancer in dogs and humans that include prostate, skin, mammary, lymphoid neoplasia, and others [121–124]. Nonetheless, it should be recognized that just as in other models, both similarities and dissimilarities exist [119], including disparities concerning genomic factors, clinical behavior, and prevalence. For example, the BRAF gene's somatic mutation occurs in nearly 60% of melanoma from humans, but only in approximately 6% of dogs [125]. Also, while osteosarcoma most typically affects the appendicular skeleton and metastasizes to lungs in humans and dogs, peak onset occurs at a young age in humans, but more often at an advanced age in dogs [126].

Stem cell research is a recent and increasing field for canines, unraveling the development of novel cell-based disease models, drug discovery, and therapies. Some research with ASCs has been performed dogs due to their regeneration properties. Canine MSCs (cMSCs), for example, successfully recovered damaged spinal cord neurons [127], increased tubular epithelial cell proliferation in cisplatin-induced kidney damage [128], successfully treated osteonecrosis [129], repaired infarcted myocardial tissue [130], are capable of chondrogenic differentiation [131] and suppression of inflammation of ruptured crucial ligament [132].

As previously discussed, ASCs cells have limitations when considered for therapy or regenerative medicine, such as limited proliferation, expansion, and differentiation potentials. In contrast, pluripotent stem cells can fill a critical void in regenerative medicine by allowing autologous studies or gene editing for in vivo or in vitro disease modeling. Similar to pigs, but in particular in dogs, isolation of genuinely pluripotent cells has been challenging. According to [133] six studies from 2007 until 2009 derived ESCs from blastocysts that expressed the core pluripotency markers and were capable of differentiating into representative lineages of all three germ layers in vitro; however, a limited proliferative potential and differentiation in germ cells layers were observed in all [134–139]. Moreover, a consensus regarding typical morphology and cell culture conditions are still unreported [136, 138].

iPSCs generation in canines has evolved quickly. Some studies have shown that the generation of canine induced pluripotency stem cells (ciPSCs) from fetal or adult cells through retroviral transduction of dog, human, or mouse factors [140–149]. The pluripotency state of the ciPSCs (naïve, primed, or other) has been discussed, and the proper characterization of these cells lacks consensus. These studies tested different medium and supplement combinations in culture and reported different pluripotency acquisition requirements and maintenance (different culture supplementation and different cell surface markers detection). Interestingly, [141] obtained ciPSCs derived from adipose multipotent stromal cells that showed similarity to human ESCs regarding morphology, pluripotency markers expression, and the ability to differentiate into all three derivatives germ layers in vitro (endoderm, ectoderm, and mesoderm).

Remarkably, dogs develop breed-associated genetic predispositions to particular disorders and suffer from many of the same maladies as humans. Many genetic diseases, such as Alzheimer's disease, retinal atrophy, muscular dystrophy, cancer, obesity, cardiovascular diseases, and diabetes mellitus, affect dogs and humans [121, 135, 150]. For instance, the neurobehavioral syndrome called canine cognitive dysfunction (CCD), which affects 14.2–22.5% of dogs over eight years old, shares many clinical and neuropathological similarities with human aging and early stages of Alzheimer's Disease [151–154]. Recently, Hyttel and collaborators [155] aimed to characterize the CCD condition in iPSC-derived neurons from aged demented and healthy dogs, allowing the comparison of CCD with human Alzheimer's at the cellular level. Canine iPSCs have also been tested in other studies, as researchers transplanted autologous iPSCs into the myocardial wall of dogs to examine the potential for myocardial infarct treatment, and the stem cell population were tracked regarding distribution, migration, engraftment, survival, proliferation, and differentiation [142].

Although biotechnological techniques and tools for the dog are less developed than for other species such as swine, the progress on gene editing technologies that can correct genetic defects, thereby offering potential treatment of some inherited diseases, is of great interest in canines due to the genetic proximity to humans described before [156, 157]. In 2015, [158] explored the feasibility of producing gene knockout (KO) dogs using gene editing by CRISPR/CAS9. The study focused to knock out the myostatin gene (MSTN), that is a negative regulator of skeletal muscle mass and demonstrated for the first time that a single injection of Cas9 mRNA and sgRNA corresponding to a particular gene into zygotes, combined with an embryo transfer strategy, efficiently generated site-specific genome-modified dogs [158, 159].

Recent studies also focused on using CRISPR/Cas9 edition for canine cancer models [160, 161]. Eun et al. [161], reported the attempt to optimize the CRISPR/Cas9 system to target canine tumor protein 53 (TP53), one of the most important tumor suppressor genes. The establishment of TP53 knockout canine cells could generate a useful platform to reveal novel oncogenic functions and effects of developing anti-cancer therapeutics [161].

Whereas one of the key benefits of using ciPSCs in disease modeling is the already discussed advantage over murine models, mostly due to the higher similarity between dogs to humans, another important perspective about ciPSCs is its potential use in clinical applications to improve the health and welfare of dogs themselves, an important aspect to be considered, in particular, due to the increasing inclusion of pets inside families and their overall importance to the One Health concept.

3. Future perspectives and final considerations

Herein advantages and hurdles of using of induced pluripotent stem cells were discussed concerning ongoing and future applications in large animal models, summarized in **Figure 1**. While true ESCs have only been described in mouse and rat models, it is widely accepted that these models are not the most adequate for studies on cellular therapies in regenerative medicine. Therefore, progress on translational medicine relies on the development of pluripotent-based technologies in suitable environments such as NH-primates, swine, or canine organisms. The alliance between in vitro induced pluripotency and gene editing tools opens a new road to suitable and experimental preclinical protocols. Besides the in vivo or in vitro disease modeling, the validation of pluripotency in domestic and wild animals holds great promise to contribute to animal production, preservation, and health by enabling, for example, the generation of gene editing and improved gametes, embryos, and animals.

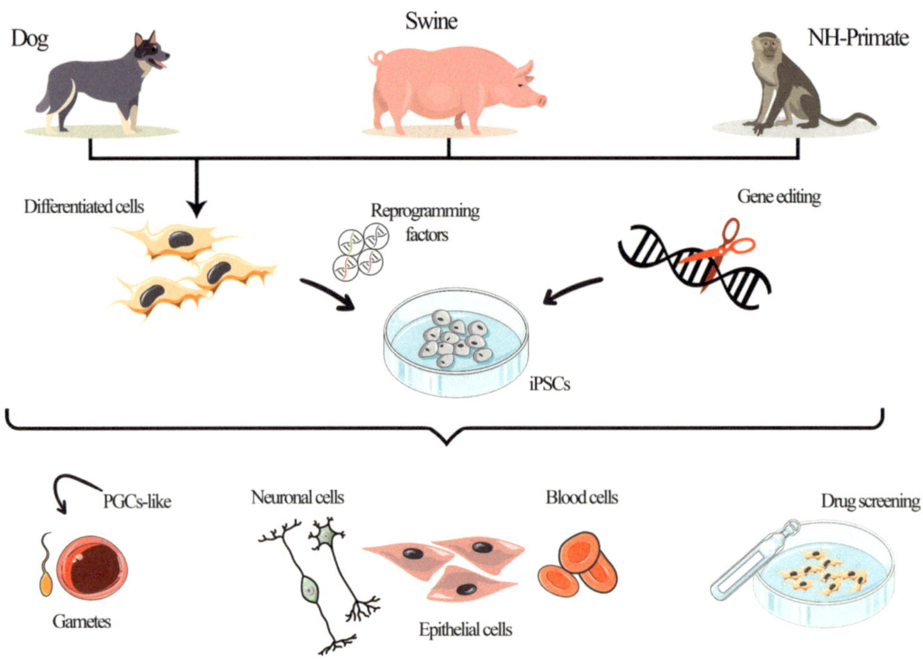

Figure 1.
Biomedical and regenerative possibilities for translational use of induced pluripotent stem cells derived from large animal models.

Acknowledgements

Funding was provided by FAPESP (São Paulo Research Foundation) grants n°. 2013/08135-2 and 2015/26818-5.

Author details

Laís Vicari de Figueiredo Pessôa[1], Naira Caroline Godoy Pieri[1], Kaiana Recchia[1,2] and Fabiana Fernandes Bressan[1,2]*

1 Faculty of Animal Science and Food Engineering, University of São Paulo, Pirassununga, São Paulo, Brazil

2 Faculty of Veterinary Medicine and Animal Science, University of São Paulo, São Paulo, Brazil

*Address all correspondence to: fabianabressan@usp.br

References

[1] Ericsson AC, Crim MJ, Franklin CL. A brief history of animal modeling. Mo Med [Internet]. 110(3):201-5. Available from: http://www.ncbi.nlm.nih.gov/pubmed/23829102

[2] ANDERSEN ML, WINTER LMF. Animal models in biological and biomedical research - experimental and ethical concerns. An Acad Bras Cienc [Internet]. 2019;91(suppl 1). Available from: http://www.scielo.br/scielo.php?script=sci_arttext&pid=S0001-37652019000200701&tlng=en

[3] Chan AWS, Cheng P-H, Neumann A, Yang J-J. Reprogramming Huntington Monkey Skin Cells into Pluripotent Stem Cells. Cell Reprogram [Internet]. 2010 Oct;**12**(5):509-517 Available from: http://www.liebertpub.com/doi/10.1089/cell.2010.0019

[4] Friedman H, Ator N, Haigwood N, Newsome W, Allen JS, Golos TG, et al. The Critical Role of Nonhuman Primates in Medical Research - White Paper. Pathog Immun [Internet]. 2017 Aug 21;2(3):352. Available from: http://paijournal.com/index.php/paijournal/article/view/186

[5] Harding JD. Nonhuman Primates and Translational Research: Progress, Opportunities, and Challenges. ILAR J [Internet]. 2017 Dec 1;**58**(2):141-150 Available from: https://academic.oup.com/ilarjournal/article/58/2/141/4745719

[6] Daadi MM, Barberi T, Shi Q, Lanford RE. Nonhuman Primate Models in Translational Regenerative Medicine. Stem Cells Dev [Internet]. 2014 Dec;**23**(S1):83-87 Available from: https://www.liebertpub.com/doi/10.1089/scd.2014.0374

[7] Cox LA, Comuzzie AG, Havill LM, Karere GM, Spradling KD, Mahaney MC, et al. Baboons as a model to study genetics and epigenetics of human disease. ILAR J [Internet]. 2013;54(2):106-21. Available from: http://www.ncbi.nlm.nih.gov/pubmed/24174436

[8] Mora-Bermúdez F, Badsha F, Kanton S, Camp JG, Vernot B, Köhler K, et al. Differences and similarities between human and chimpanzee neural progenitors during cerebral cortex development. Elife [Internet]. 2016;5. Available from: http://www.ncbi.nlm.nih.gov/pubmed/27669147

[9] Camacho P, Fan H, Liu Z, He J-Q. Large Mammalian Animal Models of Heart Disease. J Cardiovasc Dev Dis [Internet]. 2016 Oct 5;3(4). Available from: http://www.ncbi.nlm.nih.gov/pubmed/29367573

[10] Yang G, Hong H, Torres A, Malloy KE, Choudhury GR, Kim J, et al. Standards for Deriving Nonhuman Primate-Induced Pluripotent Stem Cells, Neural Stem Cells and Dopaminergic Lineage. Int J Mol Sci [Internet]. 2018 Sep 17;19(9). Available from: http://www.ncbi.nlm.nih.gov/pubmed/30227600

[11] Dudley DM, Aliota MT, Mohr EL, Weiler AM, Lehrer-Brey G, Weisgrau KL, et al. A rhesus macaque model of Asian-lineage Zika virus infection. Nat Commun [Internet]. 2016 Nov 28;7(1):12204. Available from: http://www.nature.com/articles/ncomms12204

[12] Seo J, Lee Y, Kim BS, Park J, Yang S, Yoon H-J, et al. A non-human primate model for stable chronic Parkinson's disease induced by MPTP administration based on individual behavioral quantification. J Neurosci Methods [Internet]. 2019 Jan;311:277-87. Available from: https://linkinghub.elsevier.com/retrieve/pii/S0165027018303510

[13] Hessell AJ, Jaworski JP, Epson E, Matsuda K, Pandey S, Kahl C, et al. Early short-term treatment with neutralizing human monoclonal antibodies halts SHIV infection in infant macaques. Nat Med [Internet]. 2016 Apr 21;**22**(4):362-368 Available from: http://www.nature.com/articles/nm.4063

[14] Völter CJ, Mundry R, Call J, Seed AM. Chimpanzees flexibly update working memory contents and show susceptibility to distraction in the self-ordered search task. Proc R Soc B Biol Sci [Internet]. 2019 Jul 24;286(1907):20190715. Available from: https://royalsocietypublishing.org/doi/10.1098/rspb.2019.0715

[15] Li T, Ai Z, Ji W. Primate stem cells: bridge the translation from basic research to clinic application. Sci China Life Sci [Internet]. 2019 Jan 9;**62**(1):12-21 Available from: http://link.springer.com/10.1007/s11427-018-9334-2

[16] Thomson JA, Kalishman J, Golos TG, Durning M, Harris CP, Becker RA, et al. Isolation of a primate embryonic stem cell line. Proc Natl Acad Sci [Internet]. 1995 Aug 15;**92**(17):7844-7848 Available from: http://www.pnas.org/cgi/doi/10.1073/pnas.92.17.7844

[17] Branco E, Miranda CMFC, Lima AR, Silva KSM, Cabral RM, Miranda MS, et al. Bone marrow mononuclear cells versus mesenchymal stem cells from adipose tissue on bone healing in an Old World primate: can this be extrapolated to humans? Arq Bras Med Veterinária e Zootec [Internet]. 2019 Jun;**71**(3):917-928 Available from: http://www.scielo.br/scielo.php?script=sci_arttext&pid=S0102-09352019000300917&tlng=en

[18] Pogozhykh O, Pogozhykh D, Neehus A-L, Hoffmann A, Blasczyk R, Müller T. Molecular and cellular characteristics of human and non-human primate multipotent stromal cells from the amnion and bone marrow during long term culture. Stem Cell Res Ther [Internet]. 2015 Dec 22;6(1):150. Available from: http://stemcellres.com/content/6/1/150

[19] Devine SM, Bartholomew AM, Mahmud N, Nelson M, Patil S, Hardy W, et al. Mesenchymal stem cells are capable of homing to the bone marrow of non-human primates following systemic infusion. Exp Hematol [Internet]. 2001 Feb;**29**(2):244-255 Available from: https://linkinghub.elsevier.com/retrieve/pii/S0301472X00006354

[20] Sasaki E, Hanazawa K, Kurita R, Akatsuka A, Yoshizaki T, Ishii H, et al. Establishment of Novel Embryonic Stem Cell Lines Derived from the Common Marmoset (Callithrix jacchus). Stem Cells [Internet]. 2005 Oct;**23**(9):1304-1313 Available from: http://doi.wiley.com/10.1634/stemcells.2004-0366

[21] Suemori H, Tada T, Torii R, Hosoi Y, Kobayashi K, Imahie H, et al. Establishment of embryonic stem cell lines from cynomolgus monkey blastocysts produced by IVF or ICSI. Dev Dyn [Internet]. 2001 Oct;222(2):273-279. Available from: http://doi.wiley.com/10.1002/dvdy.1191

[22] Shimozawa N, Nakamura S, Takahashi I, Hatori M, Sankai T. Characterization of a novel embryonic stem cell line from an ICSI-derived blastocyst in the African green monkey. REPRODUCTION [Internet]. 2010 Mar;**139**(3):565-573 Available from: https://rep.bioscientifica.com/view/journals/rep/139/3/565.xml

[23] Simerly CR, Navara CS, Castro CA, Turpin JC, Redinger CJ, Mich-Basso JD, et al. Establishment and characterization of baboon embryonic stem cell lines: An Old World Primate model for regeneration and transplantation research. Stem Cell Res [Internet]. 2009 May;**2**(3):178-187

Available from: https://linkinghub.elsevier.com/retrieve/pii/S1873506109000221

[24] Takahashi K, Yamanaka S. Induction of Pluripotent Stem Cells from Mouse Embryonic and Adult Fibroblast Cultures by Defined Factors. 2006;2:663-76.

[25] Grow DA, McCarrey JR, Navara CS. Advantages of nonhuman primates as preclinical models for evaluating stem cell-based therapies for Parkinson's disease. Stem Cell Res [Internet]. 2016 Sep;**17**(2):352-366 Available from: https://linkinghub.elsevier.com/retrieve/pii/S187350611630112X

[26] Pessôa LV de F, Bressan FF, Freude KK. Induced pluripotent stem cells throughout the animal kingdom: Availability and applications. World J Stem Cells [Internet]. 2019 Aug 26;**11**(8):491-505 Available from: https://www.wjgnet.com/1948-0210/full/v11/i8/491.htm

[27] Liu H, Zhu F, Yong J, Zhang P, Hou P, Li H, et al. Correspondence Generation of Induced Pluripotent Stem Cells from Adult Rhesus Monkey Fibroblasts. Stem Cell [Internet]. 2008;**3**(6):587-590 Available from: http://dx.doi.org/10.1016/j.stem.2008.10.014

[28] Ben-Nun IF, Montague SC, Houck ML, Tran HT, Garitaonandia I, Leonardo TR, et al. Induced pluripotent stem cells from highly endangered species. Nat Methods [Internet]. 2011 Sep 4;8(10):829-31. Available from: http://www.ncbi.nlm.nih.gov/pubmed/21892153

[29] Shimozawa N, Ono R, Shimada M, Shibata H, Takahashi I, Inada H, et al. Cynomolgus monkey induced pluripotent stem cells established by using exogenous genes derived from the same monkey species. Differentiation [Internet]. 85(4-5):131-9. Available from: http://www.ncbi.nlm.nih.gov/pubmed/23792767

[30] Domingues S, Masson Y, Marteyn A, Allouche J, Perrier AL, Peschanski M, et al. Differentiation of nonhuman primate pluripotent stem cells into functional keratinocytes. Stem Cell Res Ther [Internet]. 2017 Dec 19;8(1):285. Available from: https://stemcellres.biomedcentral.com/articles/10.1186/s13287-017-0741-9

[31] Tomioka I, Maeda T, Shimada H, Kawai K, Okada Y, Igarashi H, et al. Generating induced pluripotent stem cells from common marmoset (Callithrix jacchus) fetal liver cells using defined factors, including Lin28. Genes Cells [Internet]. 2010 Sep 1;15(9):959-69. Available from: http://www.ncbi.nlm.nih.gov/pubmed/20670273

[32] Navara CS, Hornecker J, Grow D, Chaudhari S, Hornsby PJ, Ichida JK, et al. Derivation of induced pluripotent stem cells from the baboon: a nonhuman primate model for preclinical testing of stem cell therapies. Cell Reprogram [Internet]. 2013 Dec;15(6):495-502. Available from: http://www.ncbi.nlm.nih.gov/pubmed/24182315

[33] Ramaswamy K, Yik WY, Wang X-M, Oliphant EN, Lu W, Shibata D, et al. Derivation of induced pluripotent stem cells from orangutan skin fibroblasts. BMC Res Notes [Internet]. 2015 Oct 16;8:577. Available from: http://www.ncbi.nlm.nih.gov/pubmed/26475477

[34] Nakai R, Ohnuki M, Kuroki K, Ito H, Hirai H, Kitajima R, et al. Derivation of induced pluripotent stem cells in Japanese macaque (*Macaca fuscata*). Sci Rep [Internet]. 2018 Dec 15;8(1):12187. Available from: http://www.nature.com/articles/s41598-018-30734-w

[35] Thoma EC, Heckel T, Keller D, Giroud N, Leonard B, Christensen K, et al. Establishment of a translational

endothelial cell model using directed differentiation of induced pluripotent stem cells from Cynomolgus monkey. Sci Rep [Internet]. 2016;6:35830. Available from: http://www.ncbi.nlm.nih.gov/pubmed/27779219

[36] Zhang X, Cao H, Bai S, Huo W, Ma Y. Differentiation and characterization of rhesus monkey atrial and ventricular cardiomyocytes from induced pluripotent stem cells. Stem Cell Res [Internet]. 2017 Apr;20:21-9. Available from: https://linkinghub.elsevier.com/retrieve/pii/S1873506117300156

[37] D'Souza SS, Maufort J, Kumar A, Zhang J, Smuga-Otto K, Thomson JA, et al. GSK3β Inhibition Promotes Efficient Myeloid and Lymphoid Hematopoiesis from Non-human Primate-Induced Pluripotent Stem Cells. Stem Cell Reports [Internet]. 2016 Feb;**6**(2):243-256 Available from: https://linkinghub.elsevier.com/retrieve/pii/S2213671115003756

[38] Hallett PJ, Deleidi M, Astradsson A, Smith GA, Cooper O, Osborn TM, et al. Successful Function of Autologous iPSC-Derived Dopamine Neurons following Transplantation in a Non-Human Primate Model of Parkinson's Disease. Cell Stem Cell [Internet]. 2015 Mar;**16**(3):269-274 Available from: https://linkinghub.elsevier.com/retrieve/pii/S1934590915000569

[39] Aron Badin R, Bugi A, Williams S, Vadori M, Michael M, Jan C, et al. MHC matching fails to prevent long-term rejection of iPSC-derived neurons in non-human primates. Nat Commun [Internet]. 2019 Dec 25;10(1):4357. Available from: http://www.nature.com/articles/s41467-019-12324-0

[40] Cho IK, Yang B, Forest C, Qian L, Chan AWS. Amelioration of Huntington's disease phenotype in astrocytes derived from iPSC-derived neural progenitor cells of Huntington's disease monkeys. PLoS One [Internet]. 2019;14(3):e0214156. Available from: http://www.ncbi.nlm.nih.gov/pubmed/30897183

[41] Vermilyea SC, Babinski A, Tran N, To S, Guthrie S, Kluss JH, et al. In Vitro CRISPR/Cas9-Directed Gene Editing to Model LRRK2 G2019S Parkinson's Disease in Common Marmosets. Sci Rep [Internet]. 2020 Feb 26;10(1):3447. Available from: http://www.ncbi.nlm.nih.gov/pubmed/32103062

[42] Kawamura T, Miyagawa S, Fukushima S, Maeda A, Kashiyama N, Kawamura A, et al. Cardiomyocytes Derived from MHC-Homozygous Induced Pluripotent Stem Cells Exhibit Reduced Allogeneic Immunogenicity in MHC-Matched Non-human Primates. Stem Cell Reports [Internet]. 2016 Mar;**6**(3):312-320 Available from: https://linkinghub.elsevier.com/retrieve/pii/S2213671116000291

[43] Zhao X, Chen H, Xiao D, Yang H, Itzhaki I, Qin X, et al. Comparison of Non-human Primate versus Human Induced Pluripotent Stem Cell-Derived Cardiomyocytes for Treatment of Myocardial Infarction. Stem Cell Reports [Internet]. 2018 Feb;**10**(2):422-435 Available from: https://linkinghub.elsevier.com/retrieve/pii/S2213671118300286

[44] Sosa E, Chen D, Rojas EJ, Hennebold JD, Peters KA, Wu Z, et al. Differentiation of primate primordial germ cell-like cells following transplantation into the adult gonadal niche. Nat Commun [Internet]. 2018 Dec 17;9(1):5339. Available from: http://www.nature.com/articles/s41467-018-07740-7

[45] Olivier EN, Wang K, Grossman J, Mahmud N, Bouhassira EE. Differentiation of Baboon (Papio anubis) Induced-Pluripotent Stem Cells into Enucleated Red Blood Cells. Cells [Internet]. 2019 Oct 19;8(10):1282.

Available from: https://www.mdpi.com/2073-4409/8/10/1282

[46] McGill TJ, Stoddard J, Renner LM, Messaoudi I, Bharti K, Mitalipov S, et al. Allogeneic iPSC-Derived RPE Cell Graft Failure Following Transplantation Into the Subretinal Space in Nonhuman Primates. Investig Opthalmology Vis Sci [Internet]. 2018 Mar 8;59(3):1374. Available from: http://iovs.arvojournals.org/article.aspx?doi=10.1167/iovs.17-22467

[47] Shiba Y, Gomibuchi T, Seto T, Wada Y, Ichimura H, Tanaka Y, et al. Allogeneic transplantation of iPS cell-derived cardiomyocytes regenerates primate hearts. Nature [Internet]. 2016 Oct 10;**538**(7625):388-391 Available from: http://www.nature.com/articles/nature19815

[48] Kunkanjanawan T, Carter R, Ahn K-S, Yang J, Parnpai R, Chan AWS. Induced Pluripotent HD Monkey Stem Cells Derived Neural Cells for Drug Discovery. SLAS Discov Adv Sci Drug Discov [Internet]. 2017 Jul 27;**22**(6):696-705 Available from: http://journals.sagepub.com/doi/10.1177/2472555216685044

[49] Goodnight A V., Kremsky I, Khampang S, Jung YH, Billingsley JM, Bosinger SE, et al. Chromatin accessibility and transcription dynamics during in vitro astrocyte differentiation of Huntington's Disease Monkey pluripotent stem cells. Epigenetics Chromatin [Internet]. 2019 Dec 13;12(1):67. Available from: https://epigeneticsandchromatin.biomedcentral.com/articles/10.1186/s13072-019-0313-6

[50] Vermilyea SC, Guthrie S, Meyer M, Smuga-Otto K, Braun K, Howden S, et al. Induced Pluripotent Stem Cell-Derived Dopaminergic Neurons from Adult Common Marmoset Fibroblasts. Stem Cells Dev [Internet]. 2017 Sep;**26**(17):1225-1235 Available from: https://www.liebertpub.com/doi/10.1089/scd.2017.0069

[51] Emborg ME, Liu Y, Xi J, Zhang X, Yin Y, Lu J, et al. Induced Pluripotent Stem Cell-Derived Neural Cells Survive and Mature in the Nonhuman Primate Brain. Cell Rep [Internet]. 2013 Mar;**3**(3):646-650 Available from: https://linkinghub.elsevier.com/retrieve/pii/S2211124713000752

[52] Hong H, Roy-Choudhury G, Kim J, Daadi MM. Isolation and Differentiation of Self-Renewable Neural Stem Cells from Marmoset-Induced Pluripotent Stem Cells. In 2019. p. 199-204. Available from: http://link.springer.com/10.1007/978-1-4939-9007-8_15

[53] Yang L, Esvelt KM, Aach J, Guell M, Dicarlo JE, Norville JE, et al. RNA-Guided Human Genome. 2013;(February):823-7.

[54] Wu M, Wei C, Lian Z, Liu R, Zhu C, Wang H, et al. Rosa26-targeted sheep gene knock-in via CRISPR-Cas9 system. Sci Rep. 2016;**6**(September 2015):1-7

[55] Cho SW, Kim S, Kim Y, Kweon J, Kim HS, Bae S, et al. Sup2. Cold Spring Harb Lab Press Method. 2014;132-41.

[56] Fu Y, Sander JD, Reyon D, Cascio VM, Joung JK. Improving CRISPR-Cas nuclease specificity using truncated guide RNAs. Nat Biotechnol. 2014;**32**(3):279-284

[57] Niu Y, Li T, Ji W. Paving the road for biomedicine: genome editing and stem cells in primates. Natl Sci Rev [Internet]. 2017 Jul 1;**4**(4):543-549 Available from: https://academic.oup.com/nsr/article/4/4/543/4093906

[58] Chen Y, Niu Y, Ji W. Genome editing in nonhuman primates: approach to generating human disease models. J Intern Med [Internet]. 2016 Sep;**280**(3):246-251 Available from: http://doi.wiley.com/10.1111/joim.12469

[59] Yang S-H, Cheng P-H, Banta H, Piotrowska-Nitsche K, Yang J-J, Cheng ECH, et al. Towards a transgenic model of Huntington's disease in a non-human primate. Nature [Internet]. 2008 Jun 18;**453**(7197):921-924 Available from: http://www.nature.com/articles/nature06975

[60] Niu Y, Guo X, Chen Y, Wang C-E, Gao J, Yang W, et al. Early Parkinson's disease symptoms in -synuclein transgenic monkeys. Hum Mol Genet [Internet]. 2015 Apr 15;**24**(8):2308-2317 Available from: https://academic.oup.com/hmg/article-lookup/doi/10.1093/hmg/ddu748

[61] Liu H, Chen Y, Niu Y, Zhang K, Kang Y, Ge W, et al. TALEN-Mediated Gene Mutagenesis in Rhesus and Cynomolgus Monkeys. Cell Stem Cell [Internet]. 2014 Mar;**14**(3):323-328 Available from: https://linkinghub.elsevier.com/retrieve/pii/S1934590914000484

[62] Chen Y, Zheng Y, Kang Y, Yang W, Niu Y, Guo X, et al. Functional disruption of the dystrophin gene in rhesus monkey using CRISPR/Cas9. Hum Mol Genet [Internet]. 2015 Jul 1;**24**(13):3764-3774 Available from: https://academic.oup.com/hmg/article-lookup/doi/10.1093/hmg/ddv120

[63] Sato K, Oiwa R, Kumita W, Henry R, Sakuma T, Ito R, et al. Generation of a Nonhuman Primate Model of Severe Combined Immunodeficiency Using Highly Efficient Genome Editing. Cell Stem Cell [Internet]. 2016 Jul;**19**(1):127-138 Available from: https://linkinghub.elsevier.com/retrieve/pii/S1934590916301539

[64] EU (2010) Directive 2010/63/EU of the European Parliament and of the Council of 22 September 2010 on the protection of animals used for scientific purposes (text with EEA relevance). OJ L 276/33.

[65] Carvalho C, Gaspar A, Knight A, Vicente L. Ethical and Scientific Pitfalls Concerning Laboratory Research with Non-Human Primates, and Possible Solutions. Animals [Internet]. 2018 Dec 29;9(1):12. Available from: http://www.mdpi.com/2076-2615/9/1/12

[66] Perleberg C, Kind A, Schnieke A. Genetically engineered pigs as models for human disease. DMM Dis Model Mech. 2018;11(1).

[67] Roberts RM, Telugu BPVL, Ezashi T. Induced pluripotent stem cells from swine (Sus scrofa): Why they may prove to be important. Cell Cycle. 2009;**8**(19):3078-3081

[68] Michael Roberts R, Yuan Y, Genovese N, Ezashi T. Livestock models for exploiting the promise of pluripotent stem cells. ILAR J. 2015;**56**(1):74-82

[69] Groenen MAM, Archibald AL, Uenishi H, Tuggle CK, Takeuchi Y, Rothschild MF, et al. Europe PMC Funders Group Europe PMC Funders Author Manuscripts Europe PMC Funders Author Manuscripts Analyses of pig genomes provide insight into porcine demography and evolution. 2012;491(7424):393-8.

[70] Schmied J, Rupa P, Garvie S, Wilkie B. Immune response phenotype of allergic versus clinically tolerant pigs in a neonatal swine model of allergy. Vet Immunol Immunopathol. 2013;**154**(1-2):17-24

[71] Thomas DJ, Husmann RJ, Villamar M, Winship TR, Buck RH, Zuckermann FA. Lactobacillus rhamnosus HN001 attenuates allergy development in a pig model. PLoS One. 2011;6(2).

[72] Turner DJ, Noble PB, Lucas MP, Mitchell HW. Decreased airway narrowing and smooth muscle contraction in hyperresponsive pigs. J Appl Physiol. 2002;93(4):1296-300.

[73] Brandler MD, Powell SC, Craig DM, Quick G, McMahon TJ, Goldberg RN, et al. A novel inhaled organic nitrate that affects pulmonary vascular tone in a piglet model of hypoxia-induced pulmonary hypertension. Pediatr Res. 2005;**58**(3):531-536

[74] Davis SS, Illum L, Hinchcliffe M. Gastrointestinal transit of dosage forms in the pig. J Pharm Pharmacol. 2001;**53**(1):33-39

[75] Bassaganya-Riera J, Hontecillas R. CLA and n-3 PUFA differentially modulate clinical activity and colonic PPAR-responsive gene expression in a pig model of experimental IBD. Clin Nutr. 2006;**25**(3):454-465

[76] Nocca D, Gagner M, Abente FC, Del Genio GM, Ueda K, Assalia A, et al. Laparoscopic gastric bypass with silicone band in a pig model: Prevention of anastomotic dilatation - Feasibility study. Obes Surg. 2005;**15**(4):523-527

[77] Badin JK, Progar V, Pareddy A, Cagle J, Alloosh M, Sturek M. Effect of Age on Diabetogenicity of Alloxan in Ossabaw Miniature Swine. Comp Med. 2019;**69**(2):114-122

[78] van den Heuvel M, Sorop O, van Ditzhuijzen NS, de Vries R, van Duin RWB, Peters I, et al. The effect of bioresorbable vascular scaffold implantation on distal coronary endothelial function in dyslipidemic swine with and without diabetes. Int J Cardiol. 2018;**252**:44-51

[79] Torres-Rovira L, Astiz S, Caro A, Lopez-Bote C, Ovilo C, Pallares P, et al. Diet-induced swine model with obesity/leptin resistance for the study of metabolic syndrome and type 2 diabetes. Sci World J. 2012;**2012**

[80] Dolezalova D, Hruska-Plochan M, Bjarkam CR, Sørensen JCH, Cunningham M, Weingarten D, et al. Pig models of neurodegenerative disorders: Utilization in cell replacement-based preclinical safety and efficacy studies. J Comp Neurol. 2014;**522**(12):2784-2801

[81] Lorson MA, Spate LD, Samuel MS, Murphy CN, Lorson CL, Prather RS, et al. Disruption of the Survival Motor Neuron (SMN) gene in pigs using ssDNA. Transgenic Res. 2011;**20**(6):1293-1304

[82] Yan S, Tu Z, Liu Z, Fan N, Yang H, Yang S, et al. A huntingtin knock-in pig model recapitulates features of selective neurodegeneration in Huntington's disease. 2019;173(4):989-1002.

[83] Cai M, Shen R, Song L, Lu M, Wang J, Zhao S, et al. Bone Marrow Mesenchymal Stem Cells (BM-MSCs) Improve Heart Function in Swine Myocardial Infarction Model through Paracrine Effects. Sci Rep. 2016;6(January):1-11.

[84] Omlor GW, Lorenz S, Nerlich AG, Guehring T, Richter W. Disc cell therapy with bone-marrow-derived autologous mesenchymal stromal cells in a large porcine disc degeneration model. Eur Spine J. 2018;**27**(10):2639-2649

[85] Nichols J, Smith A. Naive and Primed Pluripotent States. Cell Stem Cell. 2009;**4**(6):487-492

[86] Hanna J, Saha K, Jaenisch R. Somatic cell reprogramming and transitions between pluripotent states: facts, hypotheses, unresolved issues. Cell. 2010;**143**(4):508-525

[87] Davidson KC, Mason EA, Pera MF. The pluripotent state in mouse and human. Development [Internet]. 2015 Sep 15;**142**(18):3090-3099 Available from: http://dev.biologists.org/cgi/doi/10.1242/dev.116061

[88] Kumari D. States of Pluripotency: Naïve and Primed Pluripotent Stem Cells. In: Pluripotent Stem Cells - From

the Bench to the Clinic [Internet]. InTech; 2016. Available from: http://www.intechopen.com/books/pluripotent-stem-cells-from-the-bench-to-the-clinic/states-of-pluripotency-na-ve-and-primed-pluripotent-stem-cells

[89] Nowak-Imialek M, Niemann H. Pluripotent cells in farm animals: state of the art and future perspectives. Reprod Fertil Dev [Internet]. 2013;25(1):103. Available from: http://www.publish.csiro.au/?paper=RD12265

[90] Du X, Feng T, Yu D, Wu Y, Zou H, Ma S, et al. Barriers for Deriving Transgene-Free Pig iPS Cells with Episomal Vectors. Stem Cells. 2015 Nov;**33**(11):3228-3238

[91] Yu J, Hu K, Smuga-otto K, Tian S, Stewart R, Igor I, et al. Human Induced Pluripotent Stem Cell Free of Vector Transgene Sequences. Science (80-). 2009;324(5928):797-801.

[92] Okita K, Matsumura Y, Sato Y, Okada A, Morizane A, Okamoto S, et al. A more efficient method to generate integration-free human iPS cells. Nat Methods. 2011;**8**(5):409-412

[93] Li D, Secher J, Hyttel P, Ivask M, Kolko M, Hall VJ, et al. Generation of transgene-free porcine intermediate type induced pluripotent stem cells. Cell Cycle. 2018;**17**(23):2547-2563

[94] Montserrat N, Bahima EG, Batlle L, Häfner S, Rodrigues AMC, González F, et al. Generation of pig iPS cells: A model for cell therapy. J Cardiovasc Transl Res. 2011;**4**(2):121-130

[95] Ao Y, Mich-Basso JD, Lin B, Yang L. High efficient differentiation of functional hepatocytes from porcine induced pluripotent stem cells. PLoS One. 2014;**9**(6):1-11

[96] Kim E, Kim M, Hwang SU, Kim J, Lee G, Park YS, et al. Neural induction of porcine-induced pluripotent stem cells and further differentiation using glioblastoma-cultured medium. J Cell Mol Med. 2019;**23**(3):2052-2063

[97] Larsen NJ, Marklund S, Kelly KA, Malek M, Tuggle CK, Yerle M, et al. New insights into porcine-human synteny conservation. Mamm Genome. 1999;**10**(5):488-491

[98] Jiang Z, He H, Hamasima N, Suzuki H, Verrinder Gibbins AM. Comparative mapping of Homo sapiens chromosome 4 (HSA4) and Sus scrofa chromosome 8 (SSC8) using orthologous genes representing different cytogenetic bands as landmarks. Genome. 2002;**45**(1):147-156

[99] Lahbib-Mansais Y, Karlskov-Mortensen P, Mompart F, Milan D, Jørgensen CB, Cirera S, et al. A high-resolution comparative map between pig chromosome 17 and human chromosomes 4, 8, and 20: Identification of synteny breakpoints. Genomics. 2005;**86**(4):405-413

[100] Cheng LT, Sun LT, Tada T. Genome editing in induced pluripotent stem cells. Genes to Cells. 2012;**17**(6):431-438

[101] Kikuchi K, Kashiwazaki N, Nagai T, Nakai M, Somfai T, Noguchi J, et al. Selected Aspects of Advanced Porcine Reproductive Technology. Reprod Domest Anim [Internet]. 2008 Jul;43:401-6. Available from: http://doi.wiley.com/10.1111/j.1439-0531.2008.01191.x

[102] Hai T, Teng F, Guo R, Li W, Zhou Q. One-step generation of knockout pigs by zygote injection of CRISPR/Cas system. Cell Res. 2014;**24**(3):372-375

[103] Whitworth KM, Lee K, Benne JA, Beaton BP, Spate LD, Murphy SL, et al. Use of the CRISPR/Cas9 System to Produce Genetically Engineered Pigs from In Vitro-Derived Oocytes and Embryos1. Biol Reprod. 2014;**91**(3):1-13

[104] Yu HH, Zhao H, Qing YB, Pan WR, Jia BY, Zhao HY, et al. Porcine zygote injection with Cas9/sgRNAResults in DMD-modified pig with muscle dystrophy. Int J Mol Sci. 2016;17(10).

[105] Zhou X, Wang L, Du Y, Xie F, Li L, Liu Y, et al. Efficient Generation of Gene-Modified Pigs Harboring Precise Orthologous Human Mutation via CRISPR/Cas9-Induced Homology-Directed Repair in Zygotes. Hum Mutat. 2016;**37**(1):110-118

[106] Li H, Yang Y, Hong W, Huang M, Wu M, Zhao X. Applications of genome editing technology in the targeted therapy of human diseases: mechanisms, advances and prospects. Signal Transduct Target Ther. 2020;5(1).

[107] Hryhorowicz M, Zeyland J, Słomski R, Lipiński D. Genetically Modified Pigs as Organ Donors for Xenotransplantation. Mol Biotechnol. 2017;**59**(9-10):435-444

[108] Wu J, Platero-Luengo A, Sakurai M, Sugawara A, Gil MA, Yamauchi T, et al. Interspecies Chimerism with Mammalian Pluripotent Stem Cells. Cell. 2017;168(3):473-486.e15.

[109] Wu J, Vilarino M, Suzuki K, Okamura D, Bogliotti YS, Park I, et al. CRISPR-Cas9 mediated one-step disabling of pancreatogenesis in pigs. Sci Rep [Internet]. 2017 Dec 5;7(1):10487. Available from: http://www.nature.com/articles/s41598-017-08596-5

[110] Goldfracht I, Efraim Y, Shinnawi R, Kovalev E, Huber I, Gepstein A, et al. Engineered heart tissue models from hiPSC-derived cardiomyocytes and cardiac ECM for disease modeling and drug testing applications. Acta Biomater. 2019;**92**:145-159

[111] Ohata K, Ott HC. Human-scale lung regeneration based on decellularized matrix scaffolds as a biologic platform. Surg Today. 2020;**50**(7):633-643

[112] Alshaikh AB, Padma AM, Dehlin M, Akouri R, Song MJ, Brännström M, et al. Decellularization and recellularization of the ovary for bioengineering applications; studies in the mouse. Reprod Biol Endocrinol [Internet]. 2020 Dec 23;18(1):75. Available from: https://rbej.biomedcentral.com/articles/10.1186/s12958-020-00630-y

[113] Bhat ZF, Kumar S, Bhat HF. In vitro meat: A future animal-free harvest. Crit Rev Food Sci Nutr. 2017;**57**(4):782-789

[114] Ostrander EA. Wayne RK. Genome Research: The canine genome; 2005

[115] Tsai KL, Clark LA, Murphy KE. Understanding hereditary diseases using the dog and human as companion model systems. Mammalian Genome. 2007.

[116] Kirk AD. Crossing the bridge: Large animal models in translational transplantation research. Immunological Reviews. 2003

[117] Horn PA, Morris JC, Neff T, Kiem HP. Stem cell gene transfer - Efficacy and safety in large animal studies. Molecular Therapy. 2004.

[118] Schneider MR, Wolf E, Braun J, Kolb HJ, Adler H. Canine embryonic stem cells: State of the art. Theriogenology. 2010;**74**(4):492-497

[119] Lucroy MD, Suckow MA. Predictive modeling for cancer drug discovery using canine models. Expert Opin Drug Discov. 2020;**15**(6):731-738

[120] Paoloni M. Khanna C. Nature Reviews Cancer: Translation of new cancer treatments from pet dogs to humans; 2008

[121] Rivera P, von Euler H. Molecular biological aspects on canine and human mammary tumors. Vet Pathol. 2011

[122] Bongiovanni L, Brachelente C, Moreno E, Welle MM. Canine epithelial skin tumours: Expression of the stem cell markers Lgr5, Lgr6 and Sox9 in light of new cancer stem cell theories. Vet Sci. 2020

[123] Liu W, Sender S, Kong W, Beck J, Sekora A, Bornemann-Kolatzki K, et al. Establishment and characterization of stable red, far-red (fR) and near infra-red (NIR) transfected canine prostate cancer cell lines. Cancer Cell Int. 2020

[124] Villarnovo D, McCleary-Wheeler AL, Richards KL. Barking up the right tree: Advancing our understanding and treatment of lymphoma with a spontaneous canine model. Current Opinion in Hematology. 2017

[125] Mochizuki H, Kennedy K, Shapiro SG, Breen MB. BRAF mutations in canine cancers. PLoS One. 2015

[126] Fenger JM, London CA, Kisseberth WC. Canine osteosarcoma: A naturally occurring disease to inform pediatric oncology. ILAR J. 2014

[127] Ryu HH, Kang BJ, Park SS, Kim Y, Sung GJ, Woo HM, et al. Comparison of mesenchymal stem cells derived from fat, bone marrow, Wharton's jelly, and umbilical cord blood for treating spinal cord injuries in dogs. Journal of Veterinary Medical Science. 2012.

[128] Lim CY, Han JI, Kim SG, Lee CM, Park HM. Evaluation of autologous bone marrow-derived mesenchymal stem cells on renal regeneration after experimentally induced acute kidney injury in dogs. Am J Vet Res. 2016;

[129] Hang D, Li F, Che W, Wu X, Wan Y, Wang J, et al. One-Stage Positron Emission Tomography and Magnetic Resonance Imaging to Assess Mesenchymal Stem Cell Survival in a Canine Model of Intervertebral Disc Degeneration. Stem Cells Dev. 2017;

[130] Vela DC, Silva GV, Assad JAR, Sousa ALS, Coulter S, Fernandes MR, et al. Histopathological study of healing after allogenic mesenchymal stem cell delivery in myocardial infarction in dogs. J Histochem Cytochem. 2009

[131] Reich CM, Raabe O, Wenisch S, Bridger PS, Kramer M, Arnhold S. Isolation, culture and chondrogenic differentiation of canine adipose tissue- and bone marrow-derived mesenchymal stem cells-A comparative study. Vet Res Commun. 2012;

[132] Muir P, Hans EC, Racette M, Volstad N, Sample SJ, Heaton C, et al. Autologous bone marrow-derived mesenchymal stem cells modulate molecular markers of inflammation in dogs with cruciate ligament rupture. PLoS One. 2016;

[133] Betts DH, Tobias IC. Canine pluripotent stem cells: Are they ready for clinical applications? Frontiers in Veterinary Science. 2015

[134] Hatoya S, Torii R, Kondo Y, Okuno T, Kobayashi K, Wijewardana V, et al. Isolation and characterization of embryonic stem-like cells from canine blastocysts. Mol Reprod Dev. 2006;

[135] Schneider MR, Wolf E, Braun J, Kolb HJ, Adler H. Canine embryo-derived stem cells and models for human diseases. Hum Mol Genet. 2008;

[136] Hayes B, Fagerlie SR, Ramakrishnan A, Baran S, Harkey M, Graf L, et al. Derivation, characterization, and in vitro differentiation of canine embryonic stem cells. Stem Cells. 2008;**26**(2):465-473

[137] Vaags AK, Rosic-Kablar S, Gartley CJ, Zheng YZ, Chesney A,

Villagómez DAF, et al. Derivation and Characterization of Canine Embryonic Stem Cell Lines with In Vitro and In Vivo Differentiation Potential. Stem Cells. 2009;

[138] Wilcox JT, Semple E, Gartley C, Brisson B a, Perrault SD, Villagómez D a F, et al. Characterization of canine embryonic stem cell lines derived from different niche microenvironments. Stem Cells Dev. 2009 Oct;18(8):1167-78.

[139] Wilcox JT, Lai JKY, Semple E, Brisson BA, Gartley C, Armstrong JN, et al. Synaptically-competent neurons derived from canine embryonic stem cells by lineage selection with EGF and noggin. PLoS One. 2011;

[140] Luo J, Cibelli JB. Conserved Role of bFGF and a Divergent Role of LIF for Pluripotency Maintenance and Survival in Canine Pluripotent Stem Cells. Stem Cells Dev. 2016;**25**(21):1670-1680

[141] Baird A, Barsby T, Guest D. Derivation of Canine Induced Pluripotent Stem Cells. Reprod Domest Anim [Internet]. 2015 Aug;**50**(4):669-676 Available from: http://doi.wiley.com/10.1111/rda.12562

[142] Lee AS, Xu D, Plews JR, Nguyen PK, Nag D, Lyons JK, et al. Preclinical Derivation and Imaging of Autologously Transplanted Canine Induced Pluripotent Stem Cells. J Biol Chem [Internet]. 2011 Sep 16;**286**(37):32697-32704 Available from: http://www.jbc.org/lookup/doi/10.1074/jbc.M111.235739

[143] Gonçalves NJN, Bressan FF, Roballo KCS, Meirelles FV, Xavier PLP, Fukumasu H, et al. Generation of LIF-independent induced pluripotent stem cells from canine fetal fibroblasts. Theriogenology [Internet]. 2017 Apr;92:75-82. Available from: https://linkinghub.elsevier.com/retrieve/pii/S0093691X17300249

[144] Koh S, Thomas R, Tsai S, Bischoff S, Lim J-H, Breen M, et al. Growth Requirements and Chromosomal Instability of Induced Pluripotent Stem Cells Generated from Adult Canine Fibroblasts. Stem Cells Dev [Internet]. 2013 Mar 15;**22**(6):951-963 Available from: https://www.liebertpub.com/doi/10.1089/scd.2012.0393

[145] Shimada H, Nakada A, Hashimoto Y, Shigeno K, Shionoya Y, Nakamura T. Generation of canine induced pluripotent stem cells by retroviral transduction and chemical inhibitors. Mol Reprod Dev [Internet]. 2009 Nov 4;**77**(1):2-2 Available from: http://doi.wiley.com/10.1002/mrd.21117

[146] Luo J, Suhr ST, Chang EA, Wang K, Ross PJ, Nelson LL, et al. Generation of leukemia inhibitory factor and basic fibroblast growth factor-dependent induced pluripotent stem cells from canine adult somatic cells. Stem Cells Dev. 2011;

[147] Whitworth DJ, Ovchinnikov D. a, Wolvetang EJ. Generation and characterization of LIF-dependent canine induced pluripotent stem cells from adult dermal fibroblasts. Stem Cells Dev. 2012 Aug;**21**(12):2288-2297

[148] Nishimura T, Hatoya S, Kanegi R, Sugiura K, Wijewardana V, Kuwamura M, et al. Generation of functional platelets from canine induced pluripotent stem cells. Stem Cells Dev. 2013;

[149] Shimada H, Nakada A, Hashimoto Y, Shigeno K, Shionoya Y, Nakamura T. Generation of canine-induced pluripotent stem cells by retroviral transduction and chemical inhibitors. Molecular Reproduction and Development. 2010.

[150] Lindblad-Toh K, Wade CM, Mikkelsen TS, Karlsson EK, Jaffe DB, Kamal M, et al. Genome sequence,

comparative analysis and haplotype structure of the domestic dog. Nature. 2005;

[151] Studzinski CM, Araujo JA, Milgram NW. The canine model of human cognitive aging and dementia: Pharmacological validity of the model for assessment of human cognitive-enhancing drugs. Progress in Neuro-Psychopharmacology and Biological Psychiatry. 2005

[152] Cotman CW, Head E. The canine (dog) model of human aging and disease: Dietary, environmental and immunotherapy approaches. J Alzheimer's Dis. 2008;**15**(4):685-707

[153] Azkona G, García-Belenguer S, Chacón G, Rosado B, León M, Palacio J. Prevalence and risk factors of behavioural changes associated with age-related cognitive impairment in geriatric dogs: PAPER. J Small Anim Pract. 2009

[154] Salvin HE, McGreevy PD, Sachdev PS, Valenzuela MJ. Under diagnosis of canine cognitive dysfunction: A cross-sectional survey of older companion dogs. Vet J. 2010

[155] Hyttel P, Pessôa LV de F, Secher JBM, Dittlau KS, Freude K, Hall VJ, et al. Oocytes, embryos and pluripotent stem cells from a biomedical perspective. Anim Reprod. 2019;16(3):508-23.

[156] Cebrian-Serrano A, Stout T, Dinnyes A. Veterinary applications of induced pluripotent stem cells: Regenerative medicine and models for disease? Veterinary Journal. 2013.

[157] Ebert AD, Liang P, Wu JC. Induced pluripotent stem cells as a disease modeling and drug screening platform. Journal of Cardiovascular Pharmacology. 2012.

[158] Zou Q, Wang X, Liu Y, Ouyang Z, Long H, Wei S, et al. Generation of gene-target dogs using CRISPR / Cas 9 system. 2015;7:580-3.

[159] McPherron AC, Lawler AM, Lee SJ. Regulation of skeletal muscle mass in mice by a new TGF-β superfamily member. Nature. 1997;

[160] Grüntzig K, Graf R, Boo G, Guscetti F, Hässig M, Axhausen KW, et al. Swiss Canine Cancer Registry 1955-2008: Occurrence of the Most Common Tumour Diagnoses and Influence of Age, Breed, Body Size, Sex and Neutering Status on Tumour Development. J Comp Pathol. 2016;

[161] Eun K, Park MG, Jeong YW, Jeong YI, Hyun S-H, Hwang WS, et al. Establishment of TP53-knockout canine cells using optimized CRIPSR/Cas9 vector system for canine cancer research. BMC Biotechnol. 2019;**19**(1):1-11

Chapter 8

Challenges for Deriving Hepatocyte-Like Cells from Umbilical Cord Mesenchymal Stem Cells for *In Vitro* Toxicology Applications

Ana Sofia Martins Serras,
Madalena Zincke dos Reis Fernandes Cipriano,
Pedro Miguel da Graça Silva and Joana Paiva Gomes Miranda

Abstract

The *in vitro* toxicology field seeks for reliable human relevant hepatic models for predicting xenobiotics metabolism and for the safety assessment of chemicals and developing drugs. The low availability and rapid loss of the phenotype or low biotransformation activity of primary hepatocytes urged the stem cell differentiation into hepatocyte-like cells (HLCs). Umbilical cord-derived mesenchymal stem cells (UC-MSC), in particular, offer a highly available cell source, with few ethical issues and higher genetic stability. However, the dynamic and complex microenvironment of liver development, including the cell-ECM and cell–cell interactions, pressure gradients (oxygen and nutrients) and growth factor signaling that are critical for the differentiation and maturation of hepatocytes, challenges the progress of *in vitro* hepatic models. Promising strategies like (i) cytokine and growth factor supplementation mimicking the liver development; (ii) epigenetic modification; and (iii) bioengineering techniques to recreate the liver microphysiological environment are gaining increasing importance for the development of relevant *in vitro* liver models to address the need for higher predictivity and cost efficiency. In this context, this chapter reviews the existing knowledge and recent advances on the approaches for deriving HLCs from UC-MSC and their application for *in vitro* toxicology.

Keywords: human neonatal mesenchymal stem cells, umbilical cord mesenchymal stem cells, hepatocyte-like cells, hepatic differentiation, liver development, epigenetic modifiers, bioengineering, *in vitro* alternative models, *in vitro* toxicology

1. Introduction

The liver is a complex organ at the anatomical and physiological level, associated with numerous vital functions, including protein and urea synthesis, and regulation of the energy metabolism. It is also the main organ responsible for xenobiotics metabolism, the reason why it is often the first to contact with their metabolic products and most of the toxins, being one of the main targets of the toxicity caused by those drugs. Indeed, drug-induced liver injury (DILI) is responsible for nearly 60% of the cases of acute liver failure [1]. Despite the increased awareness for DILI, its absolute frequency is not decreasing demonstrating the need for evaluating drugs' hepatotoxicity and for smarter *in vitro* tools to increase predictivity and to represent the patients at a population level within the drug development process [2].

Traditional *in vitro* models for hepatotoxicity studies include monolayer cell cultures (2D) and suspensions of human hepatoma cell lines or primary hepatocytes [3] (**Table 1**). The primary cultures of human hepatocytes present the most representative phenotypic and functional profile, but exhibit a short-term viability, with a quick loss of several cellular functions within the first days in culture [3], including a loss in CYP-dependent monooxygenase activities, significant downregulation of phase I and phase II enzymes, stress-related upregulation of acute-phase-response enzymes and delocalization of transporter proteins. Rat primary hepatocytes, on the other hand, have also the disadvantage of presenting interspecies differences on the biotransformation of xenobiotics [4]. To overcome such limitations, different human hepatoma cell lines have been established. These can provide a high quantity of human cells and are cost-effective. However, those benefits are often surpassed by a number of other limitations, namely their disease-like state, lower metabolizing capacity and incomplete biotransformation profile [5]. As a result, the drawbacks of the currently available models sustain the need for relevant human *in vitro* hepatotoxicity models that better resemble the *in vivo* microphysiology.

Stem cell-based hepatic models represent an important alternative to the conventional hepatic *in vitro* systems. This chapter integrates the state of the art of human umbilical cord matrix (UCM-MSCs) or blood (UCB-MSCs) hepatic differentiation and its role as an *in vitro* alternative model for biotransformation and hepatotoxicity studies.

Model	Advantages	Limitations	Ref.
Isolated hepatocytes	Obtained from whole livers or biopsies Functions close to those of hepatocytes *in vivo* Enable interspecies and pharmacogenomics studies Representative of different lobular subpopulations	Viability: 2–4 h No bile canaliculus Low availability of human tissue Interspecies differences	[6, 7]
Primary hepatocyte cultures (pHep)	Obtained from whole livers or biopsies Functions close to those of hepatocytes *in vivo* Longer viability than isolated tissue Induction/inhibition of drug-metabolizing enzymes Enable interspecies and pharmacogenomics studies	Viability: 2–4 days Early phenotypic changes Altered bile canaliculi Difficult recovery of cells and maintenance of function upon cryopreservation Low availability of human tissue Interspecies differences	[3, 7]
Hepatoma cell lines (HepG2, Huh7)	High proliferation activity and good availability Well-characterized and abundant data available	Decreased drug enzyme activities Genotype instability	[3, 5, 8, 9]

Table 1.
Advantages and limitations of traditional in vitro *models for hepatotoxicity studies.*

1.1 Umbilical cord-derived mesenchymal stem cells as an alternative stem cell source for generating hepatocyte-like cells

Recent developments in stem cell technology have paved the way for identifying novel candidate sources of cells as an attempt to increase the availability of functional human liver-like cells, as well as improving the reliability and the accuracy of drug screening *in vitro* [10, 11]. In fact, stem cells (SCs) are of human origin and possess the ability to self-replicate and differentiate into all cell types in the body. Regarding the liver, the differentiation strategies to derive hepatocyte-like cells (HLCs) from stem cells are mostly based on mimicking the development of hepatocytes *in vivo* and include the addition of soluble medium factors, the reconstruction of the cell-matrix and the intercellular interactions through the use of alternative cell culture strategies and the assessment of cell fate via genetic modifications and epigenetic modulation [12]. Yet, the major challenges on producing stem cell-derived HLCs *in vitro* are still the immature phenotype of the HLCs [13], the lack of defined endpoints of hepatic differentiation and maturation [14], the absence of relevant positive controls [15] and defining the best stem cell source.

Several approaches have been developed for deriving HLCs from human embryonic stem cells (hESCs) [7, 16, 17], induced pluripotent stem cells (iPSCs) [7, 18–20], bipotent liver progenitor cells [21] and mesenchymal stem cells (MSCs) [22–42]. hESCs display various advantages for clinical applications when compared to immortalized cell lines and primary cell cultures, since they are genetically normal (diploid) and do not possess the high donor-dependent variability observed in primary cells [43]. The use of these cells raises, however, various ethical, technical and legal concerns [44]. Other types of SCs, like iPSCs and adult stem cells, do not give rise to those ethical issues. Nevertheless, there is some evidence suggesting that iPSC therapy has the risk of leading to tumor formation, raising safety concerns that should be addressed by researchers to ensure the viability of this therapy [44]. iPSCs and hESCs also reveal a high risk for teratoma formation *in vivo* [45], exhibiting high genomic instability, through the accumulation of mutations [46], a concern that is not raised by neonatal MSCs [47]. MSCs, on the other hand, reveal many advantages over the other SCs, which make them suitable for toxicological and regenerative medicine applications. They can be isolated from non-controversial sources at a relatively low cost, do not require feeder layers and high serum conditions, reveal a satisfactory proliferative capacity *in vitro* and are less immunogenic [23, 26, 34, 48–51]. Interestingly enough, iPSC-derived MSCs have also been reported as less immunogenic [52].

MSC classification is still controversial, being commonly defined as adult, fetal or neonatal MSCs depending on its origin. Independently of their origin, MSCs are characterized according to the International Society for Cellular Therapy (ISCT) criteria [53]. The position paper published by the ISCT states that the isolated cells must display: (a) plastic adherence when maintained in standard two-dimensional (2D) culture; (b) specific surface protein expression, typically confirmed by flow cytometry where a minimum of 95% of the cell population must portray the expression of surface markers CD105, CD73 and CD90, whereas the markers CD45, CD34, CD14 or CD11b, CD79α or CD19 and HLA-DR should show less than 2% expression among the isolated cells; and (c) tri-lineage differentiation capacity of the isolated cells, that is, these cells must be shown to differentiate to osteoblasts, adipocytes and chondroblasts using standard *in vitro* tissue culture-differentiating conditions [53].

MSCs are pluripotent stem cells that can be obtained from adult tissues like adipose tissue, brain, bone marrow and pancreas [54], but can also be isolated from neonatal tissues, like fetal-blood, amniotic sac and fluid, placenta [55] and extra-embryonic tissues such as the umbilical cord (UC-MSCs) [56]. The umbilical cord, in particular, represents a source of MSCs that is readily available as it is discarded as a medical waste after birth [57]. Herein, MSCs can be isolated from the blood (UCB-MSCs) or from the matrix (UCM-MSCs) through several isolation processes, namely by enzymatic, explants or mixed enzymatic-explant digestion methods [58, 59] that result in different yields [60–62].

Depending on their origin, MSCs may present variations in morphology, proliferation potential, growth rates and differentiation capacity as well as their regenerative potential. A significant advantage of the MSCs derived from neonatal and extra-embryonic tissues over their adult counterparts is their availability, extraction using non-invasive procedures, higher isolation yields and the absence of ethical concerns [16]. Nevertheless, other advantages have been linked to those cells. Several studies have reported superior cell biological properties such as less variability resulting from the epigenetic marks related to the donor's lifestyle as well as high proliferative capacity, increased lifespan and, importantly, enhanced potency of the UC-MSCs over the other MSCs obtained from adult tissues (**Figure 1**). Indeed, in contrast to BM-MSCs, UC-MSCs maintain a significant expansion potential of 2.5 population doublings per week up to passage 22 (P22) keeping all MSC traits and genomic stability and without reaching senescence [63, 64]. Moreover, along with the adipogenic, chondrogenic and osteogenic lineages, UC-MSCs demonstrated differentiation ability into the mesodermal lineage originating from myoblasts and cardiomyocytes [65]; into the ectodermal lineage leading to neurons [66]; and into the endodermal lineage cells, giving rise to insulin-producing cells [51] and HLCs [22–42]. Besides, UC-MSCs can be obtained from donors with diverse pharmacogenetic profiles allowing for inter-individual pharmacogenomic studies and development of personalized therapies.

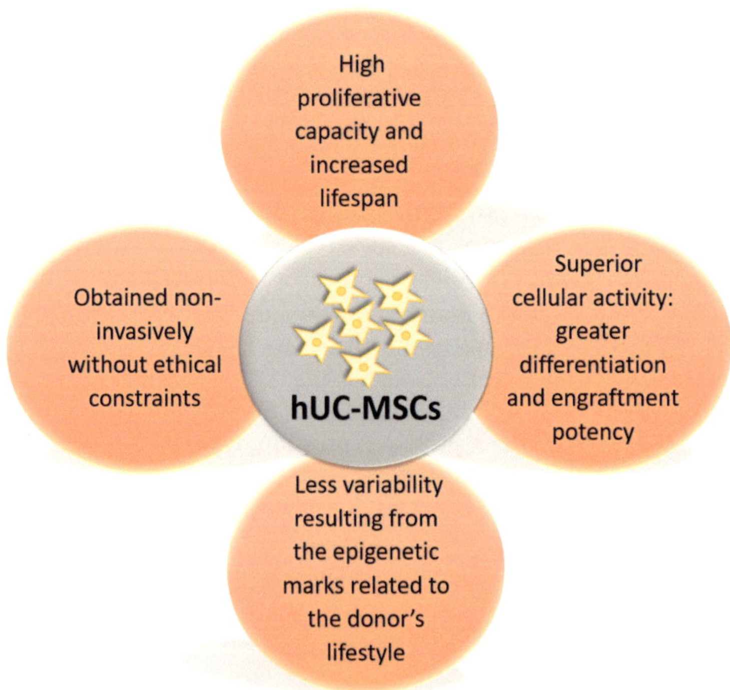

Figure 1.
Main advantages of human umbilical cord MSCs (hUC-MSCs) over other MSC sources.

2. Hepatocyte-like cell differentiation

2.1 Mimicking liver development by cytokines and growth factors supplementation

The underlying mechanisms inducing hepatocyte polarity and functional maturation *in vitro* remain largely elusive. Liver cells *in vivo* reside within a dynamic microenvironment in which biomechanical and biochemical properties of the extracellular matrix (ECM), dynamical cell-ECM and cell-cell interactions, pressure gradients (oxygen and nutrients) and growth factor signaling are critical for the differentiation and maturation of hepatocytes. The relative importance of these various factors changes during liver development and maturation. This makes developing liver models enormously challenging.

Liver development has been studied using animal models, such as mouse [67], chicken [68], zebra fish [69] and *Xenopus* [70]. The knowledge of other species' developmental biology contributed significantly to the progress and set up of protocols, which mimic the *in vivo* liver development, for deriving HLCs from human stem cells *in vitro*. As shown in **Figure 2**, the hepatogenesis process and the subsequent *in vitro* mimicking of liver development include several steps:

i. Initially, gastrulation and endoderm specification are activated by Nodal, bone morphogenetic protein (BMP) and Wnt signaling. Signaling by Nodal initiates endoderm and mesoderm formation in a concentration-dependent manner, in which high Nodal concentrations originate the definitive endoderm [71]. The endoderm induction step has been tested *in vitro* on ESCs through cell exposure to Activin A, a growth factor from the TGF-β family, which binds the same receptors as Nodal and therefore mimics its activity [72, 73].

ii. Gradients of fibroblast growth factor (FGF), Wnt, BMP and retinoic acid secreted from the adjacent mesoderm are responsible for patterning of endoderm [70, 74, 75] to generate the midgut, foregut and hindgut. Each domain

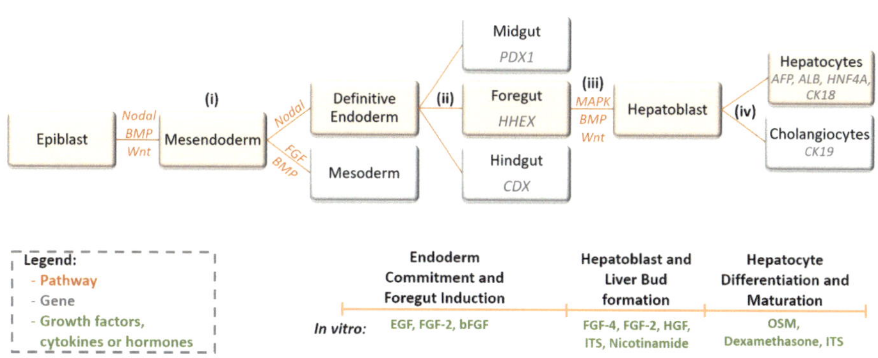

Figure 2.
Hepatogenesis and the respective inducing factors for in vitro *differentiation of MSCs into HLCs. The addition of a combination of soluble factors to the culture, at defined time points, to mimic (i) endoderm commitment and (ii) foregut induction (EGF and FGF), followed by (iii) hepatoblast and liver bud formation (FGF, HGF and ITS) and finally (iv) hepatocyte differentiation and maturation (OSM, dexamethasone and ITS) have been shown to allow hepatic differentiation to some extent, mimicking the* in vivo *ontogeny. AFP, α-fetoprotein; ALB, albumin; BMP, bone morphogenetic protein; CK, cytokeratin; EGF, epidermal growth factor; FGF, fibroblast growth factor; bFGF, basic fibroblast growth factor; HGF, hepatocyte growth factor; HNF, hepatocyte nuclear factor; ITS, insulin-transferrin-selenium; MAPK, mitogen-activated protein kinase; OSM, oncostatin M.*

expresses a specific transcription factor: HHEX in the foregut, PDX1 in the midgut and CDX in the hindgut. Activation of the *HHEX* gene expression is essential to foregut formation and therefore its development into the liver. *In vitro*, this step is mimicked by exposing cells to growth factors such as FGF and EGF, once these activate the *HHEX* gene [74, 76].

iii. After endoderm commitment and foregut induction, the foregut receives signals from the developing heart and septum transverse mesenchyme (STM), which release FGF and BMP respectively, and regulate hepatic specification to generate hepatoblasts [71, 77]. After hepatic specification, cells start expressing hepatic markers, such as α-fetoprotein (AFP), albumin (ALB), transcription factors as CEBPA and HNF4A, and change their morphology from cuboidal to pseudostratified columnar epithelium, forming the liver bud [78]. STM and hepatic mesenchyme secrete FGF, BMP, Wnt, retinoic acid and hepatocyte growth factor (HGF), which promote hepatoblast proliferation and survival [74, 75, 79, 80]. This step is generally mimicked *in vitro* using FGF to simulate the signals sent by the developing heart and STM, which induce a transformation in cell disposal and morphology [77, 79], mediating early hepatic differentiation [21]. Herein, HGF stimulates a mid-late hepatic phenotype and is commonly used to promote hepatoblast formation; however, it does not induce functional maturation [21]. FGF and HGF, as well as cell culture supplements like insulin-transferrin-sodium selenite (ITS) and nicotinamide, synergistically affect the hepatic driving pathway [34, 80].

iv. Finally, hepatoblasts, which are bipotent cells, can differentiate into hepatocytes or biliary epithelial cells. Initially, hepatoblasts express genes associated with both adult hepatocytes (*HNF4A, ALB, CK18*) and biliary epithelial cells (*CK19*), as well as fetal liver genes such as *AFP* [71]. Additionally, these cells express *CK-14, DLK1, E-cadherin, EPCAM* and *CD133* and undergo proliferation and differentiation into hepatocytes and cholangiocytes [71]. One factor responsible for the induction of hepatoblast differentiation into hepatocytes and induction of metabolic maturation is oncostatin M (OSM), secreted by hematopoietic cells in the liver [81]. Indeed, several *in vitro* models for hepatocyte differentiation use dexamethasone, HGF [22, 40, 41], OSM [25, 34, 40] and TNF-α factors [21, 71] to induce hepatocyte maturation. Moreover, the use of a collagen coating [21–23, 34] improves the *in vitro* environment to promote hepatogenic differentiation by mimicking *in vivo* ontogeny.

The protocols, to differentiate MSC into functional hepatocytes, based on the *in vivo* liver development process can be categorized into two groups: cocktail and sequential. The cocktail methodology is based on one single step, whereas the sequential and time-dependent procedures are based on four, three, or two steps. Campard et al. [22] study was one of the first described studies using a three-step-based protocol for deriving UCM-MSCs into HLCs. Herein, UCM-MSCs (also designated as Wharton's jelly cells) have been isolated by an orthodox method, involving complex vein and arterial excisions, and the authors departed from a mixed, heterogeneous population of cells. Nevertheless, after the differentiation procedure, HLC derived from UCM-MSCs exhibited a hepatocyte-like morphology, the presence of several hepatic markers (CK18, ALB, AFP and connexin 32), had glycogen storage ability, produced urea and revealed an inducible CYP3A4 activity. Still, the absence of some hepatic markers in the differentiated UC-MSCs, such as HepPar1 or HNF4A, suggested that a fully mature hepatocyte phenotype was not achieved. In another study, Zhao et al. [37], after hepatic differentiation of UC-MSC, prompted by a two-step protocol with HGF and bFGF, HLCs exhibited hepatocyte-like morphology and specific functions including albumin secretion, low-density lipoprotein uptake and urea production. In contrast, Zhang and colleagues [36], using a simpler cocktail induction protocol (with HGF and FGF-4), successfully differentiated UC-MSCs into HLCs with the same hepatic features.

Overall, those studies indicate that UC-MSCs are capable of generating hepatocyte-like cells with essential hepatic specific functions displaying an exciting potential venue toward cell-based therapeutics, human liver development studies and disease models for liver failure disorders. Yet, the weak characterization of the cells in terms of biotransformation ability has delayed their implementation for *in vitro* hepatotoxicity studies.

2.2 Epigenetic modifiers for improving HLC phenotype

Stepwise addition of factors such as EGF, FGF, HGF, nicotinamide, ITS, dexamethasone or OSM to the culture medium is used in the majority of the studies to differentiate MSCs [23, 26, 34, 35, 41]. Although the addition of these factors seems to lead to hepatic differentiation, a full mature hepatic cell has not yet been achieved. As such, the search for additional differentiation-inducing factors to induce a mature hepatic phenotype persists [34].

The normal function of cells is controlled by epigenetics, in which a combination of signaling pathways controls the balance between growth and differentiation. Therefore, besides mimicking the *in vivo* extracellular communication pathways by the use of soluble molecules, such as growth factors, cytokines, hormones and glucocorticoids, one of the strategies for controlling lineage-specific gene expression to induce a mature hepatic phenotype is by the use of chromatin remodeling agents, such as epigenetic modifiers (EM) as HDAC inhibitors (HDACi), DNMT inhibitors (DNMTi) and microRNA (miRNA).

Epigenetic modifiers change gene expression without changing the underlying DNA sequence, at the chromatin level, by modulation of its architecture between heterochromatin (transcriptionally inactive) and euchromatin (transcriptionally active) [82]. Epigenetic modulation allows to silence pluripotency transcription factors or to activate the transcription of genes of a specific lineage contributing to the improvement of the HLC phenotype [34]. These mechanisms are mainly regulated by DNA methylation, histone modifications and miRNA [82] as presented in **Figure 3**. Indeed, different strategies for hepatic differentiation based on epigenetic modification have been described so far and those include DNA methylation, histone modification and the use of microRNAs.

2.2.1 DNA methylation

DNA methyltransferases (DNMTs) introduce a methyl group, generally at CpG islands, into the DNA [82]. Decrease of DNMT1 [83] and increase of DNMT3 [84] expression have been shown to be associated with hepatic maturation. Hence, modulation of DNMTs may present a strategy for increasing liver-specific gene expression and consequently maintain a hepatic fate in HLCs [12]. 5-Azacytidine (5-AZA) is the most commonly used DNA methyltransferase inhibitor (DNMTi), whereas dimethyl sulfoxide (DMSO) emerges as a modulator of DNMTs [85].

The alteration of the DNA methylation status occurs as 5-AZA mimics the cytidine base and thus is introduced into the newly synthetized DNA strain on the S phase of the cell cycle [86]. On the subject of hepatocyte differentiation, Rothrock and colleagues [87] administered 5-AZA *in utero* to 20 days gestational age rat fetus resulting in a

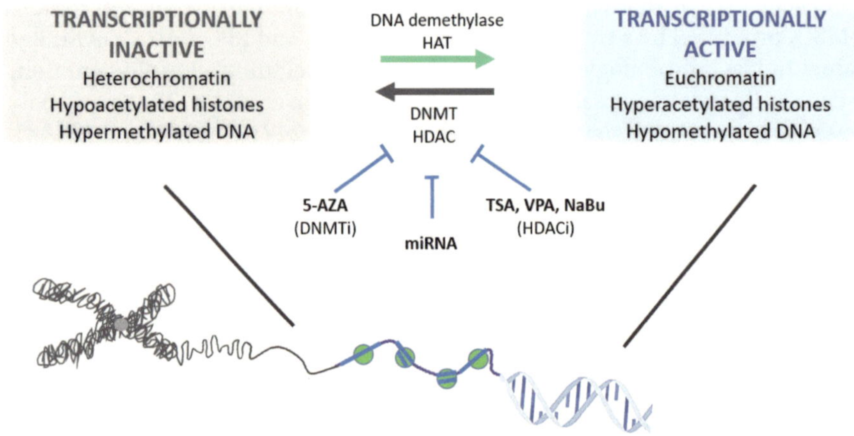

Figure 3.
Control of gene transcription by epigenetic modifiers. Hypomethylated CpG islands and hyperacetylated histone tails at the chromatin level allow gene transcription. DNMT inhibitors (DNMTi), microRNA (miRNA) and HDAC inhibitors (HDACi) modulate the chromatin structure by creating an open, transcriptionally active euchromatin. Consequently, the enhanced accessibility of transcription complexes to chromatin leads to increased transcriptional activation of several epigenetically suppressed genes. 5-AZA, 5-azacytidine; DNMT, DNA methyltransferase; HAT, histone acetyltransferase; HDAC, histone deacetylase; NaBu, sodium butyrate; TSA, trichostatin A; VPA, valproic acid.

quicker maturation of hepatocyte morphology, associated with higher activation of genes normally expressed later in liver development as *TAT* and *ALB*. *In vitro*, Yoshida et al. [25] evaluated various combinations of inducing factors, plus the addition or absence of 5-AZA, and verified that the 5-AZA combined with soluble factors was the most effective strategy for differentiating UCB-MSC that displayed a significantly higher *ALB*, *CEBPA* and *CYP1A1/2* gene expression levels, PAS positive results and urea production after 21 days of differentiation. In another study, Cipriano et al. [23] used 5-AZA as a promoter of differentiation, in the hepatoblast-like stage of differentiation, and observed a significant increase in urea production and CYP activity on HLCs.

DMSO has also been proposed as an epigenetic modifier. Although the mechanism by which DMSO induces hepatic differentiation is poorly understood, Iwatani et al. [85] suggested that it is by upregulation of the expression of DNMT3. This study was performed in mouse embryoid bodies. Yoon et al. [88], on the other hand, used trichostatin A (TSA) or DMSO treatment for the maturation steps within the hepatic differentiation procedure. Herein, TSA-treated MSCs showed higher EROD activity (human CYP450 1A1/1A2) and ammonia conversion than DMSO-treated cells. Conversely, Cipriano et al. [34] showed that, in addition to the sequential differentiation protocol, DMSO alone (in comparison to its combination with TSA) induced cellular modifications on UCM-MSCs, forming epithelial-like binucleated cells, and stimulated a homogeneous glycogen storage and improved HLC biotransformation activity. The introduction of DMSO for hepatic maturation also resulted in a significantly improved HLC phenotype and maintenance of the hepatocyte features up to 2 weeks in culture [34]. Hence, more studies are necessary in order to determine the effectiveness of DMSO on differentiation of MSCs.

2.2.2 Histone modifications

Histone deacetylases (HDCAs) remove the acetyl group from the histones and modulate chromatin to regulate cell proliferation, differentiation and growth [86]. The most commonly used histone deacetylase inhibitors (HDACi) are TSA, sodium butyrate (NaBu) and valproic acid (VPA).

Similar to other HDACi, TSA reversibly and specifically inhibits HDAC leading to hyperacetylation of histones, but the exact role in hepatic differentiation and maturation is still unclear [88]. Exposure to TSA on primary rat hepatocytes culture revealed increased cell viability and albumin secretion and maintained CYP phase I enzymes' capacity by controlling the expression of liver-enriched transcription factors (LETFs) and cell cycle arrest [89]. Yoon et al. [88] found that TSA-treated MSC presented higher activity than OSM- or DMSO-treated cells, showing an epithelial-like shape right after treatment and higher urea production and ammonia removal (compared to DMSO) on day 21, as well as the expression of late hepatic markers such as *TAT* gene expression and EROD activity. Although the expression of these markers and functions indicate a mature HLC phenotype, these values are still lower than in human primary hepatocytes. Likewise, Cipriano et al. [34] tested the effect of 100, 500 and 1000 nM of TSA on UCM-MSC differentiation and observed that 1000 nM of TSA resulted in cell detachment and cell loss, whereas 100 nM did not present relevant morphological changes from a fibroblastic morphology to a more epithelial morphology during the differentiation process. Conversely, 500 nM of TSA resulted in higher EROD and UGT activities, as well as CK18 presence and epithelial morphology [34], though, as referred in the above section, TSA-treated cells could not surpass the results with DMSO-treated cells [34].

Several protocols for hepatic differentiation of SCs prime the cells with NaBu [90], in combination to Activin A [73] or bFGF and BMP4 [91], in order to induce the definitive endoderm prior to further maturation of HLC through, for instance, DMSO [92]. However, the translation of these studies to UC-MSC is still limited. To the best of our knowledge, only Panta et al. [26] showed that pre-treatment of UCM-MSCs with NaBu upregulated hepatoblast and hepatocyte markers and stimulated mature hepatic-associated functions, such as urea production, glycogen storage and G6P, CEBPA, and CYP2B6 activity, compared to non-treated differentiated cells.

Finally, VPA, an antiepileptic and anticonvulsant drug, has demonstrated to improve stem cell hepatic differentiation when administered in low doses. An et al. [27] suggested that hepatic differentiation of UC-MSC is stimulated by VPA due to upregulation of endodermal genes such as *AKT* and *ERK*. Raut and Khana [28] also verified that pre-treatment of UCM-MSC with VPA enhanced the expression of hepatocyte-specific miRNAs typically upregulated during fetal liver development, such as miR-23b cluster (miR-27b-3p, miR-24-1-5p and miR-23b-3p), miR-30a-5p, miR-26a-5p, miR-148a-3p, miR-192-5p and miR-122-5p, which contributed to a more efficient hepatic transdifferentiation.

2.2.3 microRNA (miRNA)

MicroRNAs are critical regulators during the development of liver [93]. In humans, miR-122 is the most abundant miRNA expressed in the adult liver and is known to regulate hepatocyte differentiation [28, 93].

Zhou et al. [50] validated that, besides miR-122, also miR-148a, miR-424, miR-542-5p and miR-1246 are essential for UC-MSC differentiation, given that omitting any of these five-miRNA combination prevented hepatic transdifferentiation. In addition, it was also demonstrated that HLCs transdifferentiated from those five microRNAs expressed high level of hepatic markers in only 7 days. Moreover, Khosravi et al. [94] studied the role of embryonic overexpressed miRNAs such as miR-106a, miR-574-3p and miR-451 and determined that upregulation of any of these three alone could not induce expression of hepatic genes, such as *SOX17, FOXA2, HNF4A, ALB, AFP* and *CK18*. However, the concurrent ectopic overexpression of the three miRNAs together could induce UC-MSC differentiation into functionally mature hepatocytes in an easier, faster and efficient way compared to conventional techniques [94].

In summary, these results suggest that miRNAs have a role in hepatic differentiation and can rapidly and efficiently convert stem cells into functional HLC.

2.3 Bioengineering tools for hepatic differentiation

Hepatocytes need to be exposed to the native physiology of the liver and to have cell-cell interaction similar to the *in vivo* microenvironment in order to maintain its differentiated state [95]. Engineering tools, such as microfluidics, biomaterial scaffolds and bioprinting, have enabled greater control over the cellular microenvironment and, subsequently, cell response [96]. These strategies may set the ground for producing organs or tissues on demand to be used for animal-free drug development and personalized medicine. Moreover, optimizing cell-cell interactions using different bioengineering techniques, such as 3D liver spheroids and bioprinting, would allow a better mimic of the *in vivo* physiology and thus permit to analyze cells' response to drugs and other stimuli more accurately.

Different research groups reported various hepatic differentiation protocols by resorting to bioengineering tools. **Table 2** gathers several strategies for deriving HLCs from different umbilical cord sections both in 2D and 3D systems.

Within the umbilical cord, investigators seem to prefer using UCM-MSCs over UCB-MSCs to obtain HLCs, as shown in **Table 2**. This may be explained as

MSC source	Induction factors	Controls	Functional analysis	Hepatic markers	Ref.
	Monolayer culture				
UCB	HGF, ITS, OSM, dexamethasone	MSC	LDL uptake	*ALB, AFP, CK-18, CK-19, GS, TAT HGF, c-Met, PEPCK, CPS*	[40]
	EGF, bFGF, HGF, OSM, ITS, nicotinamide, dexamethasone	Human Hep3B cell line	Albumin and urea production, LDL uptake, glycogen storage and CYP activity	*HNF4A, CYP2B6*	[41]
UCM	HGF, FGF-4	MSC	Glycogen storage, LDL uptake	*ALB, AFP, CK18*	[36]
	Rat-tail collagen type I coating; bFGF, HGF, nicotinamide, dexamethasone, OSM, ITS	MSC and freshly isolated liver cells	Glycogen storage, G6P and CYP3A4 activity, urea production	*ALB, AFP, CK-18, CK-19, Cnx-32, TAT, TDO, CYP3A4*	[22]
	Monolayer culture; hypoxia				
UCM	HGF, FGF-4, nicotinamide, dexamethasone, OSM, ITS	MSC and HepG2	Albumin and urea production, LDL uptake, glycogen storage	*ALB, AFP, HNF4, CK-18, AAT, G6P, CYP3A4*	[42]
	Monolayer culture; overexpression of TERT				
UCM	Lentiviral transfection of MSC; EGF, HGF, ITS, OSM, dexamethasone	Untransfected MSC and HepG2	Urea production, glycogen storage	*ALB, AFP, CK-18*	[35]

MSC source	Induction factors	Controls	Functional analysis	Hepatic markers	Ref.
	Monolayer culture; epigenetic modifiers				
UCM	Rat-tail collagen; EGF, FGF-2, FGF-4, HGF, nicotinamide, dexamethasone, ITS, OSM, DMSO, 5-AZA	MSC, HepG2 and human and rat primary hepatocytes	Albumin and urea production, glycogen storage, CYP and UGT activity	*CK-18, TAT, AFP, ALB, HNF4A, CEBPA, CYP1A2, CYP3A4, OATP-C, MRP-2*	[34]
UCB	HGF, OSM, FGF-2, 5-AZA	Not-treated MSC	Urea production, glycogen storage	*ALB, CEBPA, CEBPB, PEPCK, CYP1A1, CYP1A2*	[25]
UCM	EGF, bFGF, HGF, OSM, ITS, nicotinamide, dexamethasone, NaBu	MSC not treated with NaBu, HepG2 and mouse embryonic fibroblast cell line NIH3T3	Urea production, glycogen storage	*AFP, HNF3 B, ALB, CK-18, G6P, CEBPA, CYP2B6*	[26]
UC	Rat-tail type I collagen coating; dexamethasone, HGF, OSM, ITS, VPA	MSC and MSC not treated with VPA	Albumin and urea production, glycogen storage, LDL uptake	Not studied	[27]
UCM	FGF-4, HGF, dexamethasone, OSM, VPA	MSC, differentiated MSC not treated with VPA and human adult liver biopsy	Albumin and urea production, glycogen storage	*ALB, AFP, CK-18, G6P, TAT, AAT, HNF4A, CYP3A4, CYP1A1, miR-23b cluster, miR-26a-5p, miR-30a-5p, miR-122-5p, miR-148a-3p, miR-192-5p*	[28]
	3D scaffold				
UCM	Collagen/heparin coating; IGF-1, HGF, OSM dexamethasone,	MSC, HepG2; 2D culture	Albumin production, glycogen storage, G6P and CYP2B activity	*ALB, CK-18, HNF4A, G6P, c-Met, CYP2B*	[29]
UC	GEVAC; HGF, FGF-4, ITS, dexamethasone, OSM	MSC and HepG2; 2D culture	Albumin and urea production, CYP activity	*AFP, ALB, G6P, AAT, TAT, HNF4A, CYP3A4*	[30]
	3D spheroids through cell pellet				
UC	HGF, bFGF, nicotinamide, dexamethasone, OSM, ITS	Small intestinal submucosa supplement in cell pellet; HepG2	Albumin and urea production, glycogen storage, CYP activity	*ALB, HNF4A, CYP3A4*	[31]

MSC source	Induction factors	Controls	Functional analysis	Hepatic markers	Ref.
	3D spheroids through miniaturized hollow-fiber bioreactor and self-assembled suspension culture; epigenetic modifiers				
UCM	Rat-tail collagen coating; EGF, FGF-2, HGF, nicotinamide, dexamethasone, ITS, OSM, TSA, 5-AZA, DMSO	MSC and HepG2, human and rat primary hepatocytes; 2D culture	Albumin and lactate production, glycogen storage, negative glucose consumption, CYP induction and UGT activity	*AFP, ALB, CK-18, TAT, HHEX, CEBPA, HNF4A, CYP1A1, CYP1B1, CYP3A4, OATP-C, MRP-2*	[23]
	3D spheroids through hanging drop cell culture				
UCM	IGF, HGF, OSM, dexamethasone	DMEM-treated MSC and HepG2; 2D culture	Glycogen storage	CK-18, CK-19, ALB	[32]

5-AZA, 5-azacytidine; AAT, α1 anti-trypsin; AFP, α-fetoprotein; ALB, albumin; C-Met, HGF receptor; CEBP, CCAAT enhancer-binding protein; CK, cytokeratin; CM, conditioned medium; Cnx, Connexin; CPS, carbamoyl-phosphate synthase; CYP, cytochrome P450; DMSO, dimethyl sulfoxide; EGF, epidermal growth factor; FGF, fibroblast growth factor; G6P, glucose-6-phosphatase; GEVAC, gelatin-vinyl-acetate-copolymer; GS, glutamine synthetase; HGF, hepatocyte growth factor; HNF, hepatocyte nuclear factor; IGF, insulin-like growth factor; ITS, insulin-transferrin-selenium; LDL, low-density lipoprotein; miR, microRNA; MRP, multidrug resistance protein; MSC, mesenchymal stem cells; NaBu, sodium butyrate; OATP-C, organic anion-transporting polypeptide C; OSM, oncostatin M; PEPCK, phosphoenolpyruvate carboxykinase; TAT, tyrosine aminotransferase; TDO, tryptophan-2,3-dioxygenase; TPH2, tryptophan 2,3-dioxygenase; TSA, trichostatin A; UC, umbilical cord-derived mesenchymal stem cells, UCB-MSC, umbilical cord blood-derived MSC; UCM-MSC, umbilical cord matrix-derived mesenchymal stem cells; UGT, Uridine 5′-diphosphate glucuronosyltransferase; VPA, valproic acid.

Table 2.
Protocols for differentiation of human neonatal MSC into HLC.

UCM-MSCs are easier to isolate [97] and produce higher cell numbers with better proliferation capacities when compared to UCB-MSCs [98].

2.3.1 3D cultures of UC-MSC-derived HLCs

Animal models often fail to recapitulate human biology and are not appropriate to study tissue-specific mechanisms in a controlled fashion without the interference of other tissues [99]. Thus, creating a controlled human *in vitro* tissue using 3D culture techniques is a key strategy for producing reliable knowledge on drug toxicity and disease mechanisms [23, 29, 100]. Studies using HLCs differentiated from UC-MSC through 3D systems are still scarce. Nevertheless, **Figure 4** illustrates several strategies of bioengineering for producing functional HLCs from MSCs.

Cipriano et al. [23] reported that by resorting to 3D spheroid cultures, the HLCs obtained from UCM-MSCs exhibited a higher glycogen stain and CYP3A4 induction when compared to the correspondent 2D cultures (**Figure 5**). On the other hand, HLCs cultured in hollow-fiber bioreactors favored diclofenac conversion and albumin production [23], a function mostly associated with the perivenous phenotype [101] that is also regulated by the blood flow-mediated shear stress [100]. Alternatively, Ong et al. [31] observed that MSC-derived HLCs cultured as spheroids in pellet culture endorsed expression of a subset of hepatic genes (*CYP3A4* and *HNF4A*), secreted albumin and urea, stored glycogen and showed inducible CYP3A4 mRNA levels. Importantly, the culture conditions allowed stable cell anchorage, permitted the retention of ECM molecules produced by the cells, and

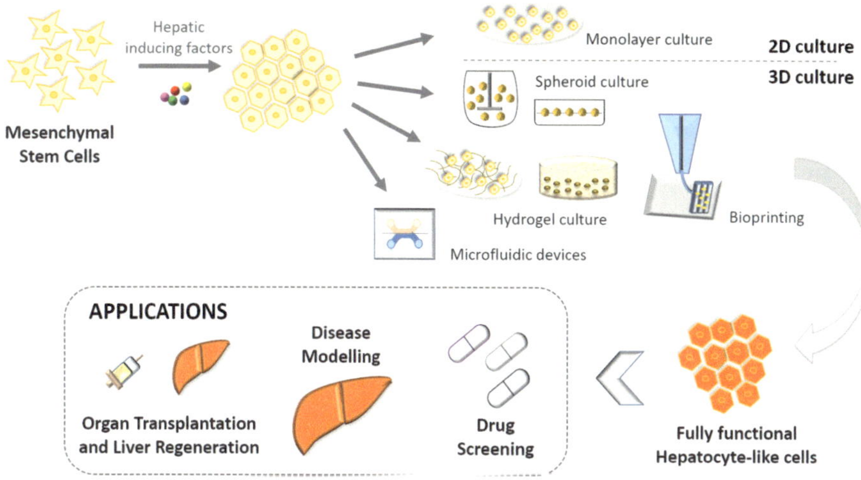

Figure 4.
2D and 3D culture strategies for the differentiation of MSCs into HLCs and potential clinical applications. Several strategies such as spheroid cultures, scaffolds, bioprinting and microfluidics have shown promissory results and represent good tools for future studies on drug screening, disease modeling and regenerative therapies using fully functional HLCs.

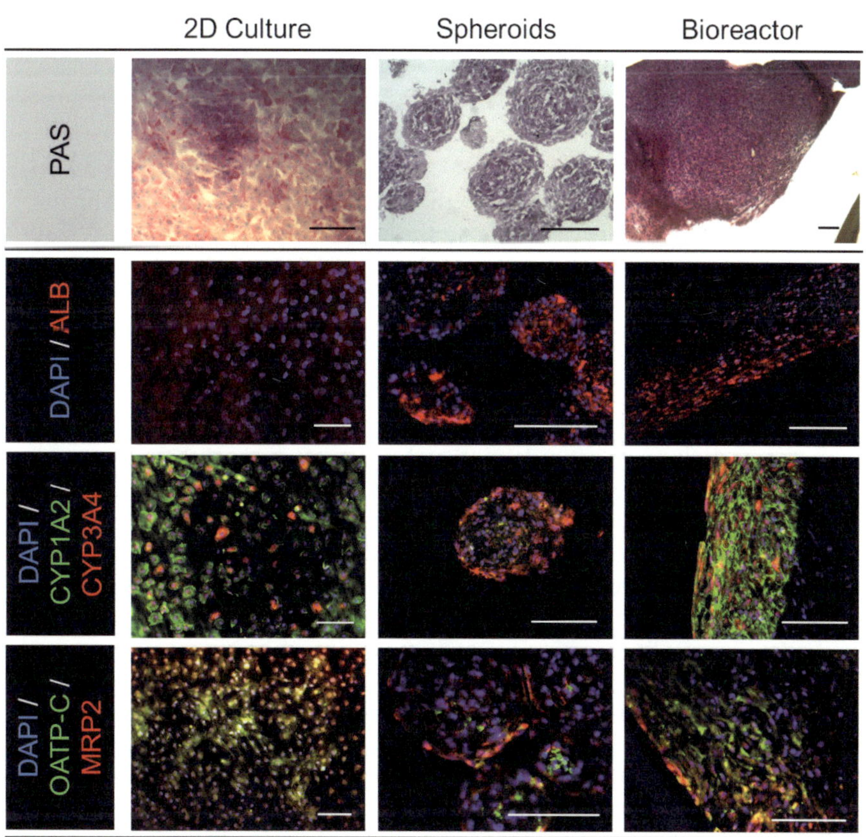

Figure 5.
Immunohistochemical and PAS analysis of HLCs-derived UCM-MSCs cultured in 2D cultures, in spheroids and in bioreactors. Representative images show the presence of the plasma protein ALB, the efflux transporter MRP-2, the uptake transporter protein OATP-C and the biotransformation enzymes CYP1A2 and CYP3A4. Cell nuclei are stained with DAPI. Scale bar: 50 μm. ALB, albumin; CYP, cytochrome P-450; HLCs, hepatocyte-like cells; MRP-2, multidrug resistance protein-2; OATP-C, organic anion-transporting polypeptide C; PAS, Periodic acid–Schiff; UCM-MSC, umbilical cord matrix mesenchymal stem cells.

when implanted into livers of hepatectomized rats also secreted human albumin into the bloodstream [31].

In the region between the blood and the hepatocytes (space of *Disse*) lies a diffuse matrix composed mostly of collagen type I and fibronectin [102]. Collagen base scaffolds have demonstrated to improve the adult hepatocyte functions as they mimic the naïve hepatocyte niche [21–23, 34]. Talaei-Khozani et al. [103] and Khodabandeh et al. [104] compared 2D and 3D cultures of UCM-MSCs in collagen films and demonstrated a better hepatogenesis and increased expression of *HNF4A* on the 3D environment. Additionally, Aleahmad et al. [29] used a 3D bioprinted collagen and heparin scaffold and verified that heparinized 2D cultures mainly expressed early liver-specific markers (e.g., *HNF4A, ALB, CK18* and *CK19*) in the presence of heparin, whereas the heparinized 3D cultures expressed both early and late liver-specific markers (e.g., *G6P, CYP2B*). In this study, HLCs showed a two-fold increase in albumin production compared to monolayer cultures [29]. These results infer that 3D culture conditions using collagen films can prevent loss of hepatocyte function and improve efficiency of hepatocyte differentiation.

Scaffolds other than collagen matrixes have also been proposed to direct hepatic differentiation of UC-MSCs. Chitrangi et al. [30] observed that gelatin-vinyl acetate (GEVAC) stimulates the differentiation of UC-MSCs into hepatospheroids, resulting in a better maturation, higher urea production, expression of CYP3A4 and CYP2C9, higher percentage of albumin-positive cells and hepatic markers compared to 2D cultures. Hashemi et al. [33] also presented a protocol for seeding UC-MSCs on a poly(ε-caprolactone) (PCL) nanofiber scaffold to stimulate and then maintain hepatic differentiation.

2.3.2 Microfluidic technologies

The hepatic zonation corresponds to the different functions revealed by the hepatocytes according to their location in the hepatic lobule, which results from the gradient of concentration of the various nutrients and oxygen observed in the hepatic environment [7]. The 3D culture systems create a gradient that may stimulate the hepatic zonation and influence HLCs obtained. For instance, the 3D configuration may mimic the liver periportal environment by generating a gradient with higher oxygen, glucose and nutrients in cells closer to the capillaries in the culture system, for example, in the outer side of spheroids, leading to higher xenobiotic metabolism, urea production and glycogen synthesis [7]. The perivenous hepatocytes have less access to oxygen supplies and are characterized by a higher xenobiotic metabolism, being exposed to physiologic conditions similar to the ones observed around the inner cells [3, 7]. Moreover, hepatocytes are not the only cells present in the liver as they interact with mesenchymal cells, stellate cells, Küpffer cells, macrophages, and lymphocytes, and are exposed, *in vivo*, to a fluid perfusion [105]. Hence, *in vitro* liver function may be optimized by resorting to microfluidic technologies.

Microfluidic culture devices (MD) permit to control the microenvironment and present the ability of continued delivery of medium, drugs and soluble molecules, allowing the study of drug metabolism and interactions [96]. The effect of medium flow on inducing albumin secretion was demonstrated by Prodanov et al. [100] using a human primary hepatocyte 3D microfluidic system. Likewise, McCarty et al. [106] demonstrated the creation of spatially-controlled zonation across multiple hepatocyte metabolism levels through the application of precise concentration gradients of exogenous hormones (insulin and glucagon) and chemical (3-methylcholanthrene) induction agents in a microfluidic device, using monolayer rat primary hepatocytes. Herein, a high concentration of insulin was directly correlated with a gradient in glycogen storage and urea production [106].

Studies reporting the hepatic differentiation of MSCs are still limited, but two studies using bone-marrow-derived MSC cultured in MDs show already a cost-effective method for HLCs production in 3 [107] to 4 weeks [108]; however, the obtained HLCs were only characterized with regard to albumin and urea quantification and showed a low metabolic performance.

3. Characterization of hepatocyte-like cells *in vitro*

A great amount of work has been developed over the past years for generating human stem cell-derived hepatocyte systems for *in vitro* toxicity testing. However, the definition of what is considered a differentiated HLC is still not unanimous and largely depends on the authors and on the purpose for which the cells are to be used. Currently, a wide and variable range of parameters is used to demonstrate the acquisition of *in vivo*-like hepatic features, which often leads to an incomplete and inconsistent cell characterization. As such, the scientific community would benefit from the harmonization and definition of the number and type of performance criteria. Indeed, Vinken and Hengstler [14] propose an optimal characterization aiming at benchmarking of hepatocyte-based *in vitro* systems for toxicity testing. This proposal comprises critical elements such as cell viability, morphology, functionality and toxicological characterization, as follows:

- Cell viability should be assessed using at least two methods that evaluate early and late key events of cytotoxicity. Moreover, a threshold of 90% viability should be adopted to discriminate between spontaneous cell death and cell death induced by toxicants.

- In terms of cellular morphology, cells must be monitored closely in order to confirm the maintenance of the hepatic polygonal shape and the detection of structural polarity markers, essential for many hepatic functions.

- Hepatocyte-specific functions, including secretion of albumin and blood coagulation factors, metabolism of carbohydrates and lipids, bile acid production and transport, as well as the detoxification of endobiotics and xenobiotics are some of the cells functionalities to be considered. This can be performed through measurement of albumin and urea production, of glycogen storage and of biotransformation enzyme activity. Other hepatic or liver-related markers, such as HNF-1/4A and PXR, should also be evaluated.

- The toxicological characterization of hepatocyte-based *in vitro* systems is important to confirm their capacity to detect prototypical types of liver toxicity. This can be achieved by using hepatotoxicants capable of replicating human *in vivo* intrinsic drug-induced liver injury, namely paracetamol dose-dependent necrosis, microvesicular steatosis induced by valproic acid and cholestasis induced by cyclosporine A.

However, a systematic interpretation of HLC-based *in vitro* systems with respect to their translation for the human *in vivo* situation remains a major challenge for future research.

4. *In vitro* toxicological applications of HLCs

Drug attrition is a major expense in the drug development process and the use of advanced *in vitro* models will likely contribute to its reduction [109]. Detection of hepatotoxicity often occurs late in the drug development process and contributes to drug attrition, withdrawn in a post-market scheme and restriction of therapeutic indications [110]. Animal-based testing is currently the base for translating *in vitro* studies to clinical trials but often do not correlate with human toxicity data [99], mainly due to interspecies differences in drug metabolism [4]. *In vitro* models cannot directly replace animal models but may occupy a new space in which, in the future, animal models will become obsolete (**Figure 6**). *In vitro* models provide tissue-specific mechanistic insights and allow to study a high number of conditions with the same cells, by means of, for example, miniaturization and higher throughput [111]. However, current *in vitro* application of differentiated HLCs in drug metabolism studies and disease modeling is still in its infancy.

Numerous studies have been published on HLCs obtained from UC-MSCs [25, 27, 34, 35, 40] as shown in **Table 2**, but few were able to demonstrate cells' biotransformation capacity, as follows: Xue et al. [49] showed that HLCs from UCM-MSCs were capable of metabolizing midazolam through CYP3A4 activity. As mentioned earlier, using a three-step protocol for UCM-MSC differentiation, Campard et al. [22] obtained HLCs that expressed important hepatic features (ALB, G6P, TDO, AAT, TAT and AFP markers, glycogen accumulation and urea production) and exhibited CYP3A4 activity. However, the absence of markers of hepatic maturation such as CYP2B6 induction after exposure to phenobarbital suggested that a relevant phenotype was not fully achieved [22]. On the other hand, Cipriano et al. [34], in contrast to the results observed when comparing HepG2 and primary human hepatocytes (PHH), obtained a comparable expression of genes involved in drug transport, amino acid metabolism and proliferation of hepatocytes between UCM-MSC-derived HLCs and PHH, indicating that HLCs are a better model for drug screening than low-metabolizing cell lines. In another study, using HLCs derived from UCM-MSC in 3D spheroids culture, Cipriano et al. [23] also observed that diclofenac was effectively converted by CYP2C9 into its hepatotoxic metabolite, 4-OH-diclofenac, and was also metabolized to a lesser extent by CYP3A4 and UGTs. Furthermore,

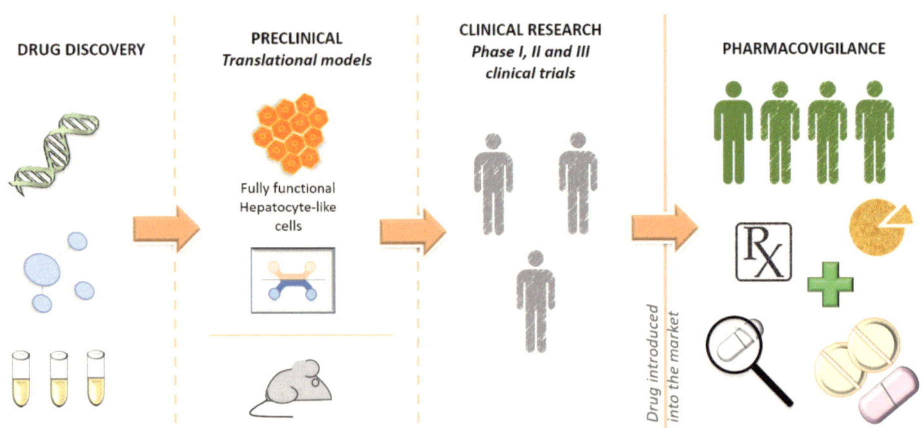

Figure 6.
Potential of mature, fully functional HLCs on drug discovery and drug development. Conventional drug development pipelines involve preclinical in vitro *and* in vivo *research in animal models followed by clinical trials on humans.* In vitro *toxicology models such as fully functional HLCs may shorten the way by performing a more similar to human toxicology screen and overcome limitations associated with poor correlation, interspecies differences and ethical concerns when using animal models.*

the determined IC50 felt in the range of what is found on the literature for primary hepatocytes, indicating that differentiated MSCs had similar dose-response characteristics to mature primary hepatocytes for this hepatotoxicant. Diclofenac is one of the most prescribed nonsteroidal anti-inflammatory drugs (NSAIDs) worldwide [112]. As such, these results permit the future use of these HLCs on drug testing and potential hepatotoxicity screening, which is often dependent of bioactivation.

Acetaminophen (APAP) is an over-the-counter antipyretic and analgesic drug widely used in several pharmacological formulations. However, in toxic doses, APAP causes liver injury by saturation of its main inactivation pathway and shifts to the transformation, by CYP2E1, CYP2D6, CYP1A2 and CYP2A6, of APAP into its hepatotoxic metabolite N-acetyl-p-benzoquinone imine (NAPQI) [113]. Chitrangi et al. [30] used hepatospheroids derived from UC-MSCs as an *in vitro* model for studying the metabolism and toxicity of APAP. CYP3A4 and CYP1A2 were induced in HLCs by APAP as well as reactive oxygen species (ROS) cell damage, which lead to cytoskeletal disorganization, both in HLCs and primary hepatocytes.

Hepatocytes derived from SCs may also represent a platform for drug discovery trough disease modeling, in which *in vivo* cell functions and mechanisms involved in pathological processes on disease onset and progression may be analyzed [114]. There are already several liver diseases successfully modeled *in vitro* using iPSC-derived hepatocytes as, for instance, familial hypercholesterolemia [115], hemophilia A [116], hepatitis C [117] and drug-induced hepatotoxicity [118]. One of the first studies to use human UC-MSCs to define a disease was Paganelli et al. [119], who developed an *in vitro* disease model to study the mechanisms underlying hepatitis B virus (HBV) infection by differentiating UC-MSCs into HLCs and infecting them with HBV. Results showed a higher susceptibility of HBV infection on HLCs rather than on undifferentiated MSCs. Despite low replication efficiency on HLCs, viral entry was as efficient as in primary hepatocytes and mimicked appropriately the *in vivo*-restricted HBV host range [119]. These similarities between the *in vivo*, *in vitro* gold standard and UC-MSCs lead, once again, to a promising opportunity for future development of *in vitro* models for drug discovery as human UC-MSCs represent a unique, human, easily available, non-transformed *in vitro* model.

5. Conclusion and future perspectives

This chapter provides insights into the potential use of human umbilical cord MSCs for obtaining a mature HLC phenotype suitable for *in vitro* toxicological studies. As primary hepatocytes present limited capacity to expand *ex vivo*, the possibility of obtaining comparable hepatocyte-like cells from MSCs may alleviate the low cell availability of human primary hepatocytes. Moreover, the use of MSCs in a physiologically relevant microenvironment that generates fully functional HLCs would allow an integrated approach to study xenobiotics biotransformation and mechanisms of action (MoA) or toxicity (MoT).

The differentiation process of MSCs into HLCs and their potential toxicology application are still in their infancy and, in the following years, there are still major challenges to resolve before their relevant application. Firstly, improvement of the efficiency of hepatic induction *in vitro* and *in vivo* still requires further investigation on the hepatic transdifferentiation mechanisms of UC-MSCs. Secondly, the differentiation process is long and the generation and maintenance of high numbers of HLCs are still difficult to achieve. The definition of the most relevant endpoints of hepatic differentiation and maturation is of extreme importance. Thirdly, the use of relevant positive controls such as human primary hepatocytes is essential for benchmarking HLCs and its absence represents a major issue in evaluating most of the published studies.

As such, further studies will be required to allow the use of HLCs derived from UC-MSCs in the drug development process, but the strategies described in this chapter represent the first step toward the establishment of a relevant human *in vitro* hepatic model for toxicological studies. Exploring the full potential of UC-MSCs by the introduction of mechanistic models for toxicity testing, including *in vitro* disease models and hepatotoxicity models, at the pipeline of drug discovery and development will significantly reduce compound attrition rate and progressively substitute current animal models by selecting safer and more efficacious lead molecules.

Acknowledgements

This work was funded by FCT (Portugal) through the research grants TUBITAK/0003/2014 and PTDC/MED-TOX/29183/2017.

Conflict of interest

The authors declare no conflict of interest.

Appendices and Nomenclature

2D	two-dimensional
3D	three-dimensional
5-AZA	5-azacytidine
AAT	α1 anti-trypsin
AFP	α-fetoprotein
ALB	albumin
BMP	bone morphogenetic protein
CEBP	CCAAT enhancer-binding protein
CK	cytokeratin
CYP	cytochrome P450
DILI	drug-induced liver injury
DMSO	dimethyl sulfoxide
DNMTi	DNA methyltransferase inhibitors
EGF	epidermal growth factor
EROD	7-ethoxyresorufin-O-deethylase
FGF	fibroblast growth factor
G6P	glucose-6-phosphatase
GS	glutamine synthetase
HDACi	histone deacetylase inhibitors
hESCs	human embryonic stem cells
HGF	hepatocyte growth factor
HLC	hepatocyte-like cell
HLCs	hepatocyte-like cells
HNF	hepatocyte nuclear factor
hnMSC	human neonatal mesenchymal stem cell
hnMSCs	human neonatal mesenchymal stem cells
hUCB-MSCs	human umbilical cord blood mesenchymal stem cells
hUCM-MSCs	human umbilical cord matrix mesenchymal stem cells
hUC-MSC	human umbilical cord-derived mesenchymal stem cell

IGF	insulin-like growth factor
iPSCs	induced pluripotent stem cells
ITS	insulin-transferrin-selenium
LDL	low-density lipoprotein
MD	microfluidic culture device
MiR	microRNA
MSC	mesenchymal stem cells
NaBu	sodium butyrate
OSM	oncostatin M
PAS	periodic acid Schiff's
SCs	stem cells
TAT	tyrosine aminotransferase
TDO	tryptophan-2,3-dioxygenase
TSA	trichostatin A
UGT	uridine 5′-diphosphate glucuronosyltransferase
VPA	valproic acid

Author details

Ana Sofia Martins Serras[1†], Madalena Zincke dos Reis Fernandes Cipriano[1,2†], Pedro Miguel da Graça Silva[1] and Joana Paiva Gomes Miranda[1*]

1 Research Institute for Medicines (iMed.ULisboa), Faculty of Pharmacy, University of Lisbon, Lisbon, Portugal

2 Fraunhofer Institute for Interfacial Engineering and Biotechnology IGB, Stuttgart, Germany

*Address all correspondence to: jmiranda@ff.ulisboa.pt

† Both authors equally contributed to this work.

References

[1] Lee WM. Drug-induced acute liver failure. Clinics in Liver Disease. 2013;**17**(4):575-586

[2] Andrade RJ, Aithal GP, Björnsson ES, Kaplowitz N, Kullak-Ublick GA, Larrey D, et al. EASL clinical practice guidelines: Drug-induced liver injury. Journal of Hepatology. 2019;**70**(6):1222-1261

[3] Gerets HHJ, Tilmant K, Gerin B, Chanteux H, Depelchin BO, Dhalluin S, et al. Characterization of primary human hepatocytes, HepG2 cells, and HepaRG cells at the mRNA level and CYP activity in response to inducers and their predictivity for the detection of human hepatotoxins. Cell Biology and Toxicology. 2012;**28**(2):69-87

[4] Martignoni M, Groothuis GMM, de Kanter R. Species differences between mouse, rat, dog, monkey and human CYP-mediated drug metabolism, inhibition and induction. Expert Opinion on Drug Metabolism & Toxicology. 2006;**2**(6):875-894

[5] Sartipy P, Björquist P, Strehl R, Hyllner J. The application of human embryonic stem cell technologies to drug discovery. Drug Discovery Today. 2007;**12**(17-18):688-699

[6] Sahu SC. Hepatotoxicity: From Genomics to In Vitro and In Vivo Models. Wiley; 2008. 704 p

[7] Godoy P, Hewitt NJ, Albrecht U, Andersen ME, Ansari N, Bhattacharya S, et al. Recent advances in 2D and 3D in vitro systems using primary hepatocytes, alternative hepatocyte sources and non-parenchymal liver cells and their use in investigating mechanisms of hepatotoxicity, cell signaling and ADME. Archives of Toxicology. 2013;**87**:1315-1530

[8] Westerink WMA, Schoonen WGEJ. Cytochrome P450 enzyme levels in HepG2 cells and cryopreserved primary human hepatocytes and their induction in HepG2 cells. Toxicology in Vitro. 2007;**21**(8):1581-1591

[9] Westerink WMA, Schoonen WGEJ. Phase II enzyme levels in HepG2 cells and cryopreserved primary human hepatocytes and their induction in HepG2 cells. Toxicology in Vitro. 2007;**21**(8):592-602

[10] Apati A, Varga N, Berecz T, Erdei Z, Homolya L, Sarkadi B. Application of human pluripotent stem cells and pluripotent stem cell-derived cellular models for assessing drug toxicity. Expert Opinion on Drug Metabolism & Toxicology. 2019;**15**(1):61-75

[11] Kim DE, Jang M-J, Kim YR, Lee J-Y, Cho EB, Kim E, et al. Prediction of drug-induced immune-mediated hepatotoxicity using hepatocyte-like cells derived from human embryonic stem cells. Toxicology. 2017;**387**:1-9

[12] Snykers S, Henkens T, De Rop E, Vinken M, Fraczek J, De Kock J, et al. Role of epigenetics in liver-specific gene transcription, hepatocyte differentiation and stem cell reprogrammation. Journal of Hepatology. 2009;**51**(1):187-211

[13] Takayama K, Morisaki Y, Kuno S, Nagamoto Y, Harada K, Furukawa N, et al. Prediction of interindividual differences in hepatic functions and drug sensitivity by using human iPS-derived hepatocytes. Proceedings of the National Academy of Sciences of the United States of America. 2014;**111**(47):16772-16777

[14] Vinken M, Hengstler JG. Characterization of hepatocyte-based in vitro systems for reliable toxicity testing. Archives of Toxicology. 2018;**92**:2981-2986

[15] Kia R, Sison RLC, Heslop J, Kitteringham NR, Hanley N, Mills JS, et al. Stem cell-derived hepatocytes as a predictive model for drug-induced liver injury: Are we there yet? British Journal of Clinical Pharmacology. 2013;**75**(4):885-896

[16] Anzalone R, Iacono ML, Corrao S, Magno F, Loria T, Cappello F, et al. New emerging potentials for human Wharton's jelly mesenchymal stem cells: Immunological features and hepatocyte-like differentiative capacity. Stem Cells and Development. 2010;**19**:423-438

[17] Prockop DJ, Oh JY. Medical therapies with adult stem/progenitor cells (MSCs): A backward journey from dramatic results in vivo to the cellular and molecular explanations. Journal of Cellular Biochemistry. 2012;**113**(5):1460-1469

[18] Freyer N, Knöspel F, Strahl N, Amini L, Schrade P, Bachmann S, et al. Hepatic differentiation of human induced pluripotent stem cells in a perfused three-dimensional multicompartment bioreactor. BioResearch Open Access. 2016;**5**(1):235-248

[19] Sjogren A-KM, Liljevald M, Glinghammar B, Sagemark J, Li X-Q, Jonebring A, et al. Critical differences in toxicity mechanisms in induced pluripotent stem cell-derived hepatocytes, hepatic cell lines and primary hepatocytes. Archives of Toxicology. 2014;**88**(7):1427-1437

[20] Viiri LE, Rantapero T, Kiamehr M, Alexanova A, Oittinen M, Viiri K, et al. Extensive reprogramming of the nascent transcriptome during iPSC to hepatocyte differentiation. Scientific Reports. 2019;**9**:3562

[21] Snykers S, De Kock J, Rogiers V, Vanhaecke T. In vitro differentiation of embryonic and adult stem cells into hepatocytes: State of the art. Stem Cells. 2009;**27**(3):577-605

[22] Campard D, Lysy PA, Najimi M, Sokal EM. Native umbilical cord matrix stem cells express hepatic markers and differentiate into hepatocyte-like cells. Gastroenterology. 2008;**134**(3):833-848

[23] Cipriano M, Freyer N, Knöspel F, Oliveira NG, Barcia R, Cruz PE, et al. Self-assembled 3D spheroids and hollow-fibre bioreactors improve MSC-derived hepatocyte-like cell maturation in vitro. Archives of Toxicology. 2017;**91**(4):1815-1832

[24] De Kock J, Vanhaecke T, Biernaskie J, Rogiers V, Snykers S. Characterization and hepatic differentiation of skin-derived precursors from adult foreskin by sequential exposure to hepatogenic cytokines and growth factors reflecting liver development. Toxicology in Vitro. 2009;**23**(8):1522-1527

[25] Yoshida Y, Shimomura T, Sakabe T, Ishii K, Gonda K, Matsuoka S, et al. A role of Wnt/β-catenin signals in hepatic fate specification of human umbilical cord blood-derived mesenchymal stem cells. American Journal of Physiology. Gastrointestinal and Liver Physiology. 2007;**293**(5):G1089-G1098

[26] Panta W, Imsoonthornruksa S, Yoisungnern T, Suksaweang S, Ketudat-Cairns M, Parnpai R. Enhanced hepatogenic differentiation of human Wharton's jelly-derived mesenchymal stem cells by using three-step protocol. International Journal of Molecular Sciences. 2019;**20**(12):1-23

[27] An SY, Han J, Lim HJ, Park SY, Kim JH, Do BR, et al. Valproic acid promotes differentiation of hepatocyte-like cells from whole human umbilical cord-derived mesenchymal stem cells. Tissue & Cell. 2014;**46**(2):127-135

[28] Raut A, Khanna A. Enhanced expression of hepatocyte-specific microRNAs in valproic acid mediated hepatic trans-differentiation of human umbilical cord derived mesenchymal

stem cells. Experimental Cell Research. 2016;**343**(2):237-247

[29] Aleahmad F, Ebrahimi S, Salmannezhad M, Azarnia M, Jaberipour M, Hoseini M, et al. Heparin/collagen 3D scaffold accelerates hepatocyte differentiation of Wharton's jelly-derived mesenchymal stem cells. Tissue Engineering and Regenerative Medicine. 2017;**14**(4):443-452

[30] Chitrangi S, Nair P, Khanna A. Three-dimensional polymer scaffolds for enhanced differentiation of human mesenchymal stem cells to hepatocyte-like cells: A comparative study. Journal of Tissue Engineering and Regenerative Medicine. 2017;**11**(8):2359-2372

[31] Ong SY, Dai H, Leong KW. Inducing hepatic differentiation of human mesenchymal stem cells in pellet culture. Biomaterials. 2006;**27**(22):4087-4097

[32] Talaei-Khozani T, Borhani-Haghighi M, Ayatollahi M, Vojdani Z. An in vitro model for hepatocyte-like cell differentiation from Wharton's jelly derived-mesenchymal stem cells by cell-base aggregates. Gastroenterology and Hepatology from Bed to Bench. 2015;**8**(3):188-199

[33] Hashemi SM, Soleimani M, Zargarian SS, Haddadi-Asl V, Ahmadbeigi N, Soudi S, et al. In vitro differentiation of human cord blood-derived unrestricted somatic stem cells into hepatocyte-like cells on poly(ε-caprolactone) nanofiber scaffolds. Cells, Tissues, Organs. 2009;**190**:135-149

[34] Cipriano M, Correia JC, Camões SP, Oliveira NG, Cruz P, Cruz H, et al. The role of epigenetic modifiers in extended cultures of functional hepatocyte-like cells derived from human neonatal mesenchymal stem cells. Archives of Toxicology. 2017;**91**(6):2469-2489

[35] Liang X, Chen X, Yang D, Huang S, Sun G, Chen Y. Differentiation of human umbilical cord mesenchymal stem cells into hepatocyte-like cells by hTERT gene transfection in vitro. Cell Biology International. 2012;**36**:215-221

[36] Zhang YN, Lie PC, Wei X. Differentiation of mesenchymal stromal cells derived from umbilical cord Wharton's jelly into hepatocyte-like cells. Cytotherapy. 2009;**11**(5):548-558

[37] Zhao Q, Ren H, Li X, Chen Z, Zhang X, Gong W, et al. Differentiation of human umbilical cord mesenchymal stromal cells into low immunogenic hepatocyte-like cells. Cytotherapy. 2009;**11**(4):414-426

[38] Tsai PC, Fu TW, Chen YMA, Ko TL, Chen TH, Shih YH, et al. The therapeutic potential of human umbilical mesenchymal stem cells from Wharton's jelly in the treatment of rat liver fibrosis. Liver Transplantation. 2009;**15**(5):484-495

[39] Yan Y, Xu W, Qian H, Si Y, Zhu W, Cao H, et al. Mesenchymal stem cells from human umbilical cords ameliorate mouse hepatic injury in vivo. Liver International. 2009;**29**(3):356-365

[40] Hong SH, Gang EJ, Jeong JA, Ahn C, Hwang SH, Yang IH, et al. In vitro differentiation of human umbilical cord blood-derived mesenchymal stem cells into hepatocyte-like cells. Biochemical and Biophysical Research Communications. 2005;**330**(4):1153-1161

[41] Lee KD, Kuo TKC, Whang-Peng J, Chung YF, Lin CT, Chou SH, et al. In vitro hepatic differentiation of human mesenchymal stem cells. Hepatology. 2004;**40**(6):1275-1284

[42] Prasajak P, Leeanansaksiri W. Developing a new two-step protocol to generate functional hepatocytes from wharton's jelly-derived mesenchymal stem cells under hypoxic condition. Stem Cells International. 2013;**9**(29):1-10

[43] Stummann TC, Bremer S. Embryonic stem cells in safety pharmacology and toxicology. Advances in Experimental Medicine and Biology. 2012;**745**:14-25

[44] Vojnits K, Bremer S. Challenges of using pluripotent stem cells for safety assessments of substances. Toxicology. 2009;**270**:10-17

[45] Herberts CA, Kwa MSG, Hermsen HPH. Risk factors in the development of stem cell therapy. Journal of Translational Medicine. 2011;**9**:29

[46] Peterson SE, Loring JF. Genomic instability in pluripotent stem cells: Implications for clinical applications. The Journal of Biological Chemistry. 2014;**289**(8):4578-4584

[47] Kim HJ, Park J-S. Usage of human mesenchymal stem cells in cell-based therapy: Advantages and disadvantages. Development & Reproduction. 2017;**21**(1):1-10

[48] Miranda JP, Filipe E, Fernandes AS, Almeida JM, Martins JP, De La Fuente A, et al. The human umbilical cord tissue-derived MSC population UCX® promotes early motogenic effects on keratinocytes and fibroblasts and G-CSF-mediatedmobilization of BM-MSCS when transplanted in vivo. Cell Transplantation. 2015;**24**(5):865-877

[49] Xue G, Han X, Ma X, Wu H, Qin Y, Liu J, et al. Effect of microenvironment on differentiation of human umbilical cord mesenchymal stem cells into hepatocytes in vitro and in vivo. BioMed Research International. 2016;**2016**:8916534

[50] Zhou X, Cui L, Zhou X, Yang Q, Wang L, Guo G, et al. Induction of hepatocyte-like cells from human umbilical cord-derived mesenchymal stem cells by defined microRNAs. Journal of Cellular and Molecular Medicine. 2017;**21**(5):881-893

[51] Yang XF, Chen T, Ren LW, Yang L, Qi H, Li FR. Immunogenicity of insulin-producing cells derived from human umbilical cord mesenchymal stem cells. Experimental and Therapeutic Medicine. 2017;**13**(4):1456-1464

[52] Khan MA, Alanazi F, Ahmed HA, Shamma T, Kelly K, Hammad MA, et al. IPSC-derived MSC therapy induces immune tolerance and supports long-term graft survival in mouse orthotopic tracheal transplants. Stem Cell Research & Therapy. 2019;**10**(1):1-15

[53] Dominici M, Le Blanc K, Mueller I, Slaper-Cortenbach I, Marini F, Krause D, et al. Minimal criteria for defining multipotent mesenchymal stromal cells. The International Society for Cellular Therapy position statement. Cytotherapy. 2006;**8**(4):315-317

[54] da Silva ML, Chagastelles PC, Nardi NB. Mesenchymal stem cells reside in virtually all post-natal organs and tissues. Journal of Cell Science. 2006;**119**(Pt 11):2204-2213

[55] de la Torre P, Jesús Pérez-Lorenzo M, Flores AI. Human placenta-derived mesenchymal stromal cells: A review from basic research to clinical applications. In: Stromal Cells - Structure, Function, and Therapeutic Implications. IntechOpen; 2019

[56] Secunda R, Vennila R, Mohanashankar AM, Rajasundari M, Jeswanth S, Surendran R. Isolation, expansion and characterisation of mesenchymal stem cells from human bone marrow, adipose tissue, umbilical cord blood and matrix: A comparative study. Cytotechnology. 2015;**67**(5):793-807

[57] Nekanti U, Mohanty L, Venugopal P, Balasubramanian S, Totey S, Ta M. Optimization and scale-up of Wharton's jelly-derived mesenchymal stem cells for clinical applications. Stem Cell Research. 2010;**5**(3):244-254

[58] Varaa N, Azandeh S, Khodabandeh Z, Gharravi AM. Wharton's jelly

mesenchymal stem cell: Various protocols for isolation and differentiation of hepatocyte-like cells; narrative review. Iranian Journal of Medical Sciences. 2019;**44**(6):437-448

[59] Hassan G, Kasem I, Soukkarieh C, Aljamali M. A simple method to isolate and expand human umbilical cord derived mesenchymal stem cells: Using explant method and umbilical cord blood serum. International Journal of Stem Cells. 2017;**10**(2):184-192

[60] Zeddou M, Briquet A, Relic B, Josse C, Malaise MG, Gothot A, et al. The umbilical cord matrix is a better source of mesenchymal stem cells (MSC) than the umbilical cord blood. Cell Biology International. 2010;**34**(7):693-701

[61] Singh TP, Sherpa ML, Pradhan A, Singh TA. Neonatal derived mesenchymal stem cells can Be isolated from human umbilical cord Wharton's jelly but not from human umbilical cord blood. Journal of Evolution of Medical and Dental Sciences. 2019;**8**(12):849-854

[62] Ganchas Soares RI, Baptista Coelho MC, Silva Santos JM, Martins JP, Basto VA, Estilita Monteiro Da Cruz P, et al. Optimised and defined method for isolation and preservation of precursor cells from human umbilical cord. WO/2009/044379. Portugal; 2009

[63] Santos JM, Camões SP, Filipe E, Cipriano M, Barcia RN, Filipe M, et al. Three-dimensional spheroid cell culture of umbilical cord tissue-derived mesenchymal stromal cells leads to enhanced paracrine induction of wound healing. Stem Cell Research & Therapy. 2015;**6**(1):1-19

[64] Martins JP, Santos JM, Almeida JM, Filipe MA, de Almeida MVT, Almeida SCP, et al. Towards an advanced therapy medicinal product based on mesenchymal stromal cells isolated from the umbilical cord tissue: Quality and safety data. Stem Cell Research & Therapy. 2014;**5**(1):9

[65] Grabowska I, Streminska W, Janczyk-Ilach K, Machaj EK, Pojda Z, Hoser G, et al. Myogenic potential of mesenchymal stem cells - the case of adhesive fraction of human umbilical cord blood cells. Current Stem Cell Research & Therapy. 2013;**8**(1):82-90

[66] Chen S, Zhang W, Wang JM, Duan HT, Kong JH, Wang YX, et al. Differentiation of isolated human umbilical cord mesenchymal stem cells into neural stem cells. International Journal of Ophthalmology. 2016;**9**(1):41-47

[67] Jung J, Zheng M, Goldfarb M, Zaret KS. Initiation of mammalian liver development from endoderm by fibroblast growth factors. Science. 1999;**284**(5422):1998-2003

[68] Fukuda-Taira S. Hepatic induction in the avian embryo: Specificity of reactive endoderm and inductive mesoderm. Journal of Embryology and Experimental Morphology. 1981;**63**:111-125

[69] Wilkins BJ, Pack M. Zebrafish models of human liver development and disease. Comprehensive Physiology. 2013;**3**(3):1213-1230

[70] McLin VA, Rankin SA, Zorn AM. Repression of Wnt/β-catenin signaling in the anterior endoderm is essential for liver and pancreas development. Development. 2007;**134**(12):2207-2217

[71] Gordillo M, Evans T, Gouon-Evans V. Orchestrating liver development. Development. 2015;**142**(12):2094-2108

[72] Hay DC, Fletcher J, Payne C, Terrace JD, Gallagher RCJ, Snoeys J, et al. Highly efficient differentiation of hESCs to functional hepatic endoderm requires ActivinA and Wnt3a signaling.

Proceedings of the National Academy of Sciences of the United States of America. 2008;**105**(34):12301-12306

[73] Zhou M, Li P, Tan L, Qu S, Ying QL, Song H. Differentiation of mouse embryonic stem cells into hepatocytes induced by a combination of cytokines and sodium butyrate. Journal of Cellular Biochemistry. 2010;**109**(3):606-614

[74] Calmont A, Wandzioch E, Tremblay KD, Minowada G, Kaestner KH, Martin GR, et al. An FGF response pathway that mediates hepatic gene induction in embryonic endoderm cells. Developmental Cell. 2006;**11**(3):339-348

[75] Wang Z, Dollé P, Cardoso WV, Niederreither K. Retinoic acid regulates morphogenesis and patterning of posterior foregut derivatives. Developmental Biology. 2006;**297**(2):433-445

[76] Shen MM. Nodal signaling: Development roles and regulation. Development. 2007;**134**(6):1023-1034

[77] Ameri J, Ståhlberg A, Pedersen J, Johansson JK, Johannesson MM, Artner I, et al. FGF2 specifies hESC-derived definitive endoderm into foregut/midgut cell lineages in a concentration-dependent manner. Stem Cells. 2010;**28**(1):45-56

[78] Bort R, Signore M, Tremblay K, Barbera JPM, Zaret KS. Hex homeobox gene controls the transition of the endoderm to a pseudostratified, cell emergent epithelium for liver bud development. Developmental Biology. 2006;**290**(1):44-56

[79] Rossi JM, Dunn NR, Hogan BLM, Zaret KS. Distinct mesodermal signals, including BMPs from the septum, transversum mesenchyme, are required in combination for hepatogenesis from the endoderm. Genes & Development. 2001;**15**(15):1998-2009

[80] Zorn AM. Liver Development. In: StemBook. Harvard Stem Cell Institute: Cambridge (MA); 2008

[81] Kamiya A, Kinoshita T, Miyajima A. Oncostatin M and hepatocyte growth factor induce hepatic maturation via distinct signaling pathways. FEBS Letters. 2001;**492**(1-2):90-94

[82] Mortada I, Mortada R. Epigenetic changes in mesenchymal stem cells differentiation. European Journal of Medical Genetics. 2017;**61**(2):114-118

[83] Kaji K, Factor VM, Andersen JB, Durkin ME, Tomokuni A, Marquardt JU, et al. DNMT1 is a required genomic regulator for murine liver histogenesis and regeneration. Hepatology. 2016;**64**(2):582-598

[84] Lee CW, Huang WC, Huang HD, Huang YH, Ho JH, Yang MH, et al. DNA methyltransferases modulate hepatogenic lineage plasticity of mesenchymal stromal cells. Stem Cell Reports. 2017;**9**(1):247-263

[85] Iwatani M, Ikegami K, Kremenska Y, Hattori N, Tanaka S, Yagi S, et al. Dimethyl sulfoxide has an impact on epigenetic profile in mouse embryoid body. Stem Cells. 2006;**24**(11):2549-2556

[86] El-Serafi AT. Epigenetic modifiers and stem cell differentiation. Cancer Stem Cells. 2012;**8**:147-154

[87] Rothrock R, Lee KL, Isham KR, Kenney FT. Changes in hepatic differentiation following treatment of rat fetuses with 5-azacytidine. Archives of Biochemistry and Biophysics. 1988;**263**(2):237-244

[88] Yoon HH, Jung BY, Seo YK, Song KY, Park JK. In vitro hepatic differentiation of umbilical cord-derived mesenchymal stem cell. Process Biochemistry. 2010;**45**(12):1857-1864

[89] Henkens T, Papeleu P, Elaut G, Vinken M, Rogiers V,

Vanhaecke T. Trichostatin A, a critical factor in maintaining the functional differentiation of primary cultured rat hepatocytes. Toxicology and Applied Pharmacology. 2007;**218**(1):64-71

[90] Yan Y, Zhong B, Qi W, Song Y, Wu Y, Fan Y, et al. In vitro differentiation of mouse embryonic stem cells into functional hepatocytes by sodium butyrate, hepatocyte growth factor and dexamethasone under chemically defined conditions. African Journal of Biotechnology. 2011;**10**(46):9493-9500

[91] Cao J, Shang CZ, Lü LH, Qiu DC, Ren M, Chen YJ, et al. Differentiation of embryonic stem cells into hepatocytes that coexpress coagulation factors VIII and IX. Acta Pharmacologica Sinica. 2010;**31**:1478-1486

[92] Hay DC, Zhao D, Fletcher J, Hewitt ZA, McLean D, Urruticoechea-Uriguen A, et al. Efficient differentiation of hepatocytes from human embryonic stem cells exhibiting markers recapitulating liver development in vivo. Stem Cells. 2008;**26**(4):894-902

[93] Deng XG, Qiu RL, Wu YH, Li ZX, Xie P, Zhang J, et al. Overexpression of miR-122 promotes the hepatic differentiation and maturation of mouse ESCs through a miR-122/FoxA1/HNF4a-positive feedback loop. Liver International. 2014;**34**(2):281-295

[94] Khosravi M, Azarpira N, Shamdani S, Hojjat-Assari S, Naserian S, Karimi MH. Differentiation of umbilical cord derived mesenchymal stem cells to hepatocyte cells by transfection of miR-106a, miR-574-3p, and miR-451. Gene. 2018;**667**:1-9

[95] Proctor WR, Foster AJ, Vogt J, Summers C, Middleton B, Pilling MA, et al. Utility of spherical human liver microtissues for prediction of clinical drug-induced liver injury. Archives of Toxicology. 2017;**91**(8):2849-2863

[96] Underhill GH, Khetani SR. Bioengineered liver models for drug testing and cell differentiation studies. Cellular and Molecular Gastroenterology and Hepatology. 2018;**5**(3):426-439.e1

[97] Reinisch A, Bartmann C, Rohde E, Schallmoser K, Bjelic-Radisic V, Lanzer G, et al. Humanized system to propagate cord blood-derived multipotent mesenchymal stromal cells for clinical application. Regenerative Medicine. 2007 Jul;**2**(4):371-382

[98] Araújo AB, Salton GD, Furlan JM, Schneider N, Angeli MH, Laureano AM, et al. Comparison of human mesenchymal stromal cells from four neonatal tissues: Amniotic membrane, chorionic membrane, placental decidua and umbilical cord. Cytotherapy. 2017;**19**(5):577-585

[99] Olson H, Betton G, Robinson D, Thomas K, Monro A, Kolaja G, et al. Concordance of the toxicity of pharmaceuticals in humans and in animals. Regulatory Toxicology and Pharmacology. 2000;**32**(1):56-67

[100] Prodanov L, Jindal R, Bale SS, Hegde M, Mccarty WJ, Golberg I, et al. Long-term maintenance of a microfluidic 3D human liver sinusoid. Biotechnology and Bioengineering. 2016;**113**:241-246

[101] Jungermann K, Keitzmann T. Zonation of parenchymal and nonparenchymal metabolism in liver. Annual Review of Nutrition. 1996;**16**:179-203

[102] Hansen LK, Wilhelm J, Fassett JT. Regulation of hepatocyte cell cycle progression and differentiation by type I collagen structure. Current Topics in Developmental Biology. 2006;**72**(05):205-236

[103] Talaei-Khozani T, Khodabandeh Z, Jaberipour M, Hosseini A, Bahmanpour S, Vojdani Z. Comparison of hepatic nuclear factor-4 expression in two and three-dimensional culture of

Wharton's jelly-derived cells exposed to hepatogenic medium. Romanian Journal of Morphology and Embryology. 2015;**56**(4):1365-1370

[104] Khodabandeh Z, Vojdani Z, Talaei-Khozani T, Jaberipour M, Hosseini A, Bahmanpour S. Comparison of the expression of hepatic genes by human Wharton's jelly mesenchymal stem cells cultured in 2D and 3D collagen culture systems. Iranian Journal of Medical Sciences. 2016;**41**(1):28-36

[105] Kryou C, Leva V, Chatzipetrou M, Zergioti I. Bioprinting for liver transplantation. Bioengineering. 2019;**6**(4):95

[106] McCarty WJ, Usta OB, Yarmush ML. A microfabricated platform for generating physiologically-relevant hepatocyte zonation. Scientific Reports. 2016;**6**:26868

[107] Yen MH, Wu YY, Liu YS, Rimando M, Ho JHC, Lee OKS. Efficient generation of hepatic cells from mesenchymal stromal cells by an innovative bio-microfluidic cell culture device. Stem Cell Research & Therapy. 2016;**7**:120

[108] Ju X, Li D, Gao N, Shi Q, Hou H. Hepatogenic differentiation of mesenchymal stem cells using microfluidic chips. Biotechnology Journal. 2008;**3**(3):383-391

[109] Franzen N, van Harten WH, Retèl VP, Loskill P, van den Eijnden-van Raaij J, IJzerman M. Impact of organ-on-a-chip technology on pharmaceutical R&D costs. Drug Discovery Today. 2019;**24**(9):1720-1724

[110] Wilke RA, Lin DW, Roden DM, Watkins PB, Flockhart D, Zineh I, et al. Identifying genetic risk factors for serious adverse drug reactions: Current progress and challenges. Nature Reviews. Drug Discovery. 2007;**6**(11):904-916

[111] Van Den Berg A, Mummery CL, Passier R, Van der Meer AD. Personalised organs-on-chips: Functional testing for precision medicine. Lab on a Chip. 2019;**2**:198-205

[112] McGettigan P, Henry D. Use of non-steroidal anti-inflammatory drugs that elevate cardiovascular risk: An examination of sales and essential medicines lists in low-, middle-, and high-income countries. PLoS Medicine. 2013;**10**(2):e1001388

[113] Mazaleuskaya LL, Sangkuhl K, Thorn CF, Fitzgerald GA, Altman RB, Klein TE. PharmGKB summary: Pathways of acetaminophen metabolism at the therapeutic versus toxic doses. Pharmacogenetics and Genomics. 2015;**25**(8):416-426

[114] Caddeo S, Boffito M, Sartori S. Tissue engineering approaches in the design of healthy and pathological in vitro tissue models. Frontiers in Bioengineering and Biotechnology. 2017;**5**:40

[115] Cayo MA, Mallanna SK, Di Furio F, Jing R, Tolliver LB, Bures M, et al. A drug screen using human iPSC-derived hepatocyte-like cells reveals cardiac glycosides as a potential treatment for hypercholesterolemia. Cell Stem Cell. 2017;**20**(4):478-489.e5

[116] Jia B, Chen S, Zhao Z, Liu P, Cai J, Qin D, et al. Modeling of hemophilia a using patient-specific induced pluripotent stem cells derived from urine cells. Life Sciences. 2014;**108**(1):22-29

[117] Sa-ngiamsuntorn K, Wongkajornsilp A, Phanthong P, Borwornpinyo S, Kitiyanant N, Chantratita W, et al. A robust model of natural hepatitis C infection using hepatocyte-like cells derived from human induced pluripotent stem cells as a long-term host. Virology Journal. 2016;**13**(59):1-18

[118] Choudhury Y, Toh YC, Xing J, Qu Y, Poh J, Huan L, et al. Patient-specific

hepatocyte-like cells derived from induced pluripotent stem cells model pazopanib-mediated hepatotoxicity. Scientific Reports. 2017;7:41238

[119] Paganelli M, Dallmeier K, Nyabi O, Scheers I, Kabamba B, Neyts J, et al. Differentiated umbilical cord matrix stem cells as a new in vitro model to study early events during hepatitis B virus infection. Hepatology. 2013;57(1):59-69